Deer: *The Animal Answer Guide*

Deer

The Animal Answer Guide

George A. Feldhamer and William J. McShea

The Johns Hopkins University Press   Baltimore

 Published 2012
Printed in the United States of America on acid-free paper
9 8 7 6 5 4 3 2 1

The Johns Hopkins University Press
2715 North Charles Street
Baltimore, Maryland 21218-4363
www.press.jhu.edu

Library of Congress Cataloging-in-Publication Data

Feldhamer, George A.
Deer : the animal answer guide / George A. Feldhamer and William J. McShea.
p. cm.
Includes bibliographical references and index.
ISBN-13: 978-1-4214-0387-8 (hardcover : acid-free paper)
ISBN-13: 978-1-4214-0388-5 (pbk. : acid-free paper)
ISBN-10: 1-4214-0387-0 (hardcover : acid-free paper)
ISBN-10: 1-4214-0388-9 (pbk. : acid-free paper)
1. Cervidae—Miscellanea. 2. Deer—Miscellanea. I. McShea, William J. II. Title.
QL737.U55F45 2012
599.65—dc23   2011019916

A catalog record for this book is available from the British Library.

Special discounts are available for bulk purchases of this book. For more information, please contact Special Sales at 410-516-6936 or specialsales@press.jhu.edu.

The Johns Hopkins University Press uses environmentally friendly book materials, including recycled text paper that is composed of at least 30 percent post-consumer waste, whenever possible.

The weather was fine and moderate. The hunters all returned, having killed during their absence three elk, four deer, two porcupines, a fox, and a hare.

Meriwether Lewis

Contents

Acknowledgments

Many friends and colleagues throughout the years have been instrumental in our interest and enthusiasm in working on various aspects of deer biology. Feldhamer is especially grateful to Joe Chapman, who hired him fresh out of graduate school in 1977 to work at the Appalachian Environmental Lab, University of Maryland, on the wild population of introduced sika deer on Maryland's Eastern Shore. Learning about this exotic species, and its ecological interactions with native white-tailed deer in the same region, led to a broader interest, appreciation, and fascination with deer in general. Much of that work on sika deer could not have been done without the able assistance of Bruce Taliaferro. Feldhamer's work on sika deer continued long after coming to the Department of Zoology at Southern Illinois University Carbondale (SIUC), which has provided his academic home since 1984.

McShea is grateful to Georg Schwede for the opportunity to wrestle deer, which set him on a new research path; John Rappole for helping tie all the research parts together; and Chris Wemmer for encouraging his exploration of deer beyond whitetails. He thanks especially all the field staff and colleagues around the world who have pointed out the nuances of wildlife that were strange and exotic.

Several colleagues have offered useful insights and suggestions on many of the chapters, which greatly augmented this book. We are especially grateful to Terry Bowyer, Biological Sciences Department, Idaho State University, and to Clay Nielsen, Cooperative Wildlife Research Laboratory, Southern Illinois University, for their editorial comments and suggestions. The manuscript greatly benefited from the thorough editorial review and comments of Howard Cincotta. Ken Collins and Travis Mossotti of the SIUC Department of English critically reviewed an earlier draft of chapter 11. Budhan Pukazhenthi provided some little-known information on deer reproduction. Several anonymous reviewers offered very valuable comments and helped correct errors of exclusion and inclusion.

Numerous colleagues from around the world graciously allowed use of their photos in this volume, including Kaleem Ahmed, Eyal Bartov, Mario Santos Beade, Tupper Ansel Blake, Arnaud Desbiez, Donna Dewhurst, Mauricio Barbanti Duarte, Anne Gunn, M. Monirul H. Khan, Dong Lei, Rexford Lord, Chen Min, Mike Ondik, Freddie Pattiselanno, Jackie Pringle, Catherine Putman, Rory Putman, Randy Rieches, Ramiro Rodriguez, Li Sheng, JoAnne Smith and Werner Flueck, Seth Stapleton, Chad Stewart, David Tuttle, Lisa Ware, and Jiagong Zhala. We also thank Paul Shelton and members of the Illinois Department of Natural Resources for use of archive photos. Lisa Russell did the pencil drawings that grace several of the chapters. Lisa, along with Kimberly Smith, SIUC Environmental Studies Program, were instrumental in many other ways to help bring this volume to fruition.

Vincent J. Burke, executive editor at the Johns Hopkins University Press, and Jennifer Malat, editorial assistant, remained positive and encouraging throughout the entire process. Thanks also to Kathleen Capels for copyediting that greatly improved the clarity of the manuscript.

Introduction

George Feldhamer

I have taught upper-division mammalogy and game management courses for 35 years. Fortunately, my career in academia also allowed for a fair amount of research work outside the classroom, with many fond memories of projects on several species of deer—with attendant good times, perils, and pitfalls.

My graduate work was on small desert rodents, so naturally my first job with the University of Maryland involved something quite different: studying introduced Japanese sika deer in Dorchester County, along the Chesapeake Bay on Maryland's Eastern Shore. My love affair with sika deer has continued. In those early years I especially enjoyed the visits to many deer-hunting camps to measure specimens and collect tissue samples for genetic analyses or the reproductive tracts of harvested females to look for twin fetuses and certain other features there. Talking with hunters and enjoying their hospitality remains a high point of those years. Equally memorable were the long days and nights during the hunting season working at deer check-stations—now being phased out in many states—to weigh specimens, collect deer jaws to determine their age, or gather other samples.

Part of my fieldwork entailed building large wooden Stephenson box traps to capture sika deer and fit them with radiocollars, to gather data on the size of their home range and their habitat use. Dorchester County is practically at sea level, and during one of the first winters of the project, an unprecedented, massive snowstorm occurred—depositing over 4 feet of snow. This was considered a once-in-500-years event. It was so bad that the National Guard was called out and the city of Cambridge closed down for a week. But despite the almost impossible conditions, my technician Bruce Taliaferro and I made our rounds, checking the big deer traps and closing them. Within a week it rained, the snow all melted, and the tide came in. Following this "perfect storm" of watery conditions, none of our traps were ever seen again—every one of them had literally floated away. I suspect the remains are scattered on the bottom of Chesapeake Bay like an old shipwreck.

Another introduced species of deer occurs on Land Between the Lakes (LBL), a 40-mile-long peninsula, now managed by the Forest Service, in western Kentucky and Tennessee. Fallow deer were first introduced to LBL in 1920, and the population quickly expanded to become the largest in North America. By the 1980s, however, it had declined significantly, and questions arose as to why. After moving to Southern Illinois University, as part of a study on the survival of fallow deer calves on LBL, we attempted to radiocollar as many as we could find. Stepping out of the truck at the beginning of the very first evening of searching, I literally stumbled over a fallow calf near the side of the road. My first thought was "Gee, this is really going to be easy. I just hope I have enough collars." After two months and countless hours of searching, that was the only calf we ever found. An important lesson in hubris!

Another memorable incident occurred during a study of white-tailed deer in Pennsylvania to determine the effectiveness of fencing in keeping the deer off an interstate freeway. Box traps baited with apples had not worked, probably because it was an extremely warm winter and the deer had little incentive to enter the traps. We were attempting to radiocollar the whitetails to monitor their movements in the area around the freeway, so we used a dart gun to deliver a mild tranquilizer drug. We had two technicians for the project, and some of the darting took place around a residential area, so they were extremely careful not to overdose the animals. We certainly did not administer too much to any deer; in fact, it was quite the opposite. Despite the proper dosage for her size, one adult female never went down, but spent several days running around with the dart dangling from her rump. So we tried a higher dosage. She still didn't go down, but now there were two darts dangling from her rump. Thankfully, the third dart was the charm. Most people who have ever worked with deer no doubt have similar memories.

Bill McShea

I did not start out as a deer ecologist. My first work was with primates, and then I received my advanced degrees in rodent population ecology. Once I arrived at the Smithsonian Institution as a postdoctoral fellow, I was stationed at their research center in Front Royal, Virginia, then under the guidance of Director Chris Wemmer. The center had a diversity of captive deer populations, including tufted deer, Père David's deer, red muntjac, and Eld's deer. Their different mating systems and social interactions were fascinating—both for their obvious links to the environments in which they evolved, and for their contrasts with white-tailed deer. Although I was hired to study the social behavior of star-nosed moles, I found myself spending more time assisting three German graduate students, Georg Schwede, Michael Stüwe, and Stephan Holzenbein, on their studies of white-tailed deer. They seemed to be having all the fun, handling and following deer through the day and night. Within two years, I was studying the interactions between rodent and deer populations as a way to justify my "rodeo" activities. This work moved into a collaboration with John Rappole, using large deer exclosures (study areas fenced to keep deer out) to look at the interactions between deer and birds in the understory vegetation. For 25 years, ever since arriving at the center, I have maintained its annual deer captures and surveys, as well as the deer exclosures.

One great advantage of working at the Smithsonian is the opportunity to be involved in international studies. One of my early trips was to Myanmar, with Steve Monfort, to study Eld's deer in their native habitat. This led to studying other Eld's deer populations throughout Southeast Asia, and eventually to expanding this research to other deer species, both in that region and in China. My expertise on tropical deer in Asia led to opportunities to study pampas deer and marsh deer in South America. Observing the diversity of deer species around the world has helped me put the ecology of white-tailed deer into perspective. The large deer exclosures, erected back in my early deer/rodent interaction studies, became very useful in studying deer impacts on forest ecology. As co-chair (with Susana González) of the International Union for Conservation of Nature (IUCN) Deer Specialist Group, I have a primary concern in conserving the rare deer species of the world while controlling the impact of the overabundant species. It makes for an interesting perspective on both sides of conservation management.

The ecology, behavior, and life-history characteristics of the world's 50 species of deer will always fascinate biologists, wildlife managers, and the general public. While most mammal species are of passing interest to everyday audiences, deer have an especially powerful fascination for many. People seem to either love deer or hate them—sometimes both at the same time. We can think of no other wildlife species in North America that have as big an impact on the activities, finances, and interests of the general public. Numerous university and agency wildlife biologists have worked tirelessly for many years to better understand the life histories of deer. This book benefits from, and reflects, their efforts. Our hope is that this volume is useful in answering some of the many questions you may have about deer in North America and elsewhere, and perhaps may inspire continued work on this highly intriguing group of "charismatic mammalian megafauna."

Deer: *The Animal Answer Guide*

Chapter 1

Introducing Deer

What are deer?

Deer are among the most common and easily recognized large animals in the world. Everyone knows what deer in their area look like, and most of us probably have an opinion about them—either positive or negative. You probably see deer regularly, and you may even have hit one with your vehicle.

In general, deer can be defined most easily as mammals with antlers. Antlers are the single characteristic that people immediately associate with deer. Although other animals may have horns, only deer have antlers, which structurally are not the same as horns. The primary difference is that antlers are deciduous—a new set grows every year and then falls off (is cast) after the mating season (rut) ends—while horns are permanent and continue to grow from year to year. Antlers also differ from horns in that they are made of bone, are branched, and normally occur only in males—except in caribou, also known as reindeer (*Rangifer tarandus*). Rarely, a reproductive female of another species may produce antlers.

The size and shape of antlers can often be used to tell the various species of deer apart. Males lack antlers in only one species, Chinese water deer (*Hydropotes inermis*). Instead, these deer have enlarged, curved, upper canine teeth. In all other species the males possess antlers, including tufted deer (*Elaphodus cephalophus*) and muntjac (genus *Muntiacus*), both of which also have enlarged upper canines. Antlers vary in their size and complexity among species—from a single, short spike in species such as pudu (genus *Pudu*), brocket deer (genus *Mazama*), and tufted deer to large main branches with numerous projections (called points or tines) in species like

Antlers of a red deer. Horns occur in several groups of hoofed mammals (ungulates), but only deer have antlers. Photo courtesy of Rory Putman.

white-tailed deer (*Odocoileus virginianus*) and caribou. Moose (*Alces alces*) and fallow deer (*Dama dama*) have palmate (flat, spade-shaped) antlers.

Majestic trophy antlers have graced the halls of castles and fired the imagination of hunters and naturalists for thousands of years. Scientists have studied the various aspects of antler growth, structure, physiology (functioning), adaptive significance, and evolutionary history for more than 200 years. Because antlers evolved primarily for male-male combat, in order to gain access to females, it is certainly appropriate that antlers have been referred to as "bones of contention."

Besides antlers, several other characteristics help to define deer. All deer eat vegetation—that is, they are herbivores. More specifically, all deer are ruminants (plant-eating mammals that chew, swallow, regurgitate, and then re-chew their food); they have much more efficient digestion—and a structurally different digestive tract—than non-ruminant herbivores, such as horses or elephants. Deer have a forestomach, with three chambers (the rumen, the reticulum, and the omasum) where food is processed before it enters the true stomach (the abomasum). Cellulose and other components of plants are broken down in the three chambers of the forestomach by microorganisms—protozoa and bacteria. After these microbes begin to digest food in the rumen, the food may be regurgitated (brought back up to the mouth), re-chewed (commonly referred to as "chewing the cud"), and then swallowed again. Additional processing occurs in the reticulum and omasum before material enters the true stomach, where acid digestion

The lack of antlers on males and the enlarged upper canine teeth distinguish Chinese water deer from all other members of the deer family Cervidae. Drawing by Lisa Russell.

occurs (see chapter 7). Being a ruminant allows a deer to consume a large initial amount of vegetation fairly quickly. Researchers have suggested that this reduces the amount of time the animal spends in open areas, where it may be more vulnerable to predators. Deer can then retire to a more secure area to chew and process their food in relative safety. Several other mammalian families are ruminants, such as the family Bovidae, which includes sheep, goats, African antelope, and cattle.

What is the difference between bucks, stags, and bulls?

Terms related to sex and age in deer can be confusing. Male deer of most North and South American species are called bucks, while the females are referred to as does. In species such as moose and North American elk (or wapiti—*Cervus elaphus*), however, the males are known as bulls and the females as cows. In many Old World species, such as red deer (considered to be the same species as North American elk) or sika deer (*Cervus nippon*), the males are called stags and the females are hinds. Young animals are either fawns or calves. In his book *The Natural History of Deer*, Rory Putman gives an excellent account of the historical derivation of these terms, as well as other European terms applied to various age groups of deer.

How many kinds of deer are there?

There are approximately 50 species (unique kinds) of deer living today. We say "approximately" because the number of species for some groups—the South American brocket deer and the Asian muntjac, for example—is still being worked out among scientists. Although deer in different areas

Antler size and shape vary among species. ***Left*****, Brocket deer, like pudu and tufted deer, have short, spike antlers.** ***Right*****, Moose have extensive, flattened (palmate) antlers.** Drawings by Lisa Russell.

may look very similar, new species have recently been described, based on differences in the number of chromosomes they have or in the structure of their DNA (see chapter 12). DNA occurs in two of the basic components of a cell: the nucleus (the cell's command center) and the mitochondria (structures that convert energy from food into a form the cell can use). Chromosomes are in the nucleus of the cell and contain DNA. Muntjac are particularly interesting in this regard, as their chromosome numbers vary from a total of 46 in Reeve's muntjac (*Muntiacus reevesi*) to only 6 in red muntjac (*M. muntjac*)—the least number found in any mammal. Recent work based on mitochondrial DNA indicates that eastern and western groups of red deer could be considered to be two separate species. Some specialists think that Persian fallow deer (*Dama mesopotamica*) are a distinct species from fallow deer (*Dama dama*), although others do not. The genetic relationships of currently recognized, living species of deer—or species groups, in the case of muntjac and brocket deer—are shown below.

How are musk deer different from "true" deer?

Musk deer look very much like "true" deer, and they were once included within the deer family Cervidae. Taxonomists (scientists who clas-

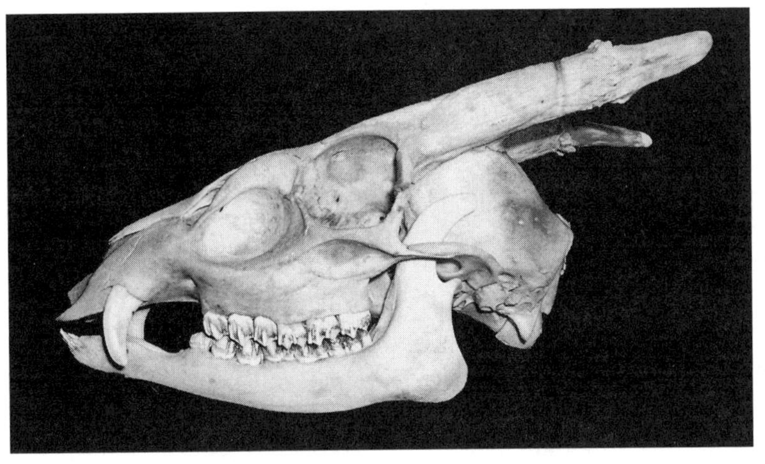

Muntjac are a group of Old World deer. *Left*, Muntjac skulls have very long bases (pedicels) for the antlers, as well as enlarged upper canine teeth. *Right*, Reeve's muntjac. Photos by George Feldhamer (*left*) and Sarefo (*right*).

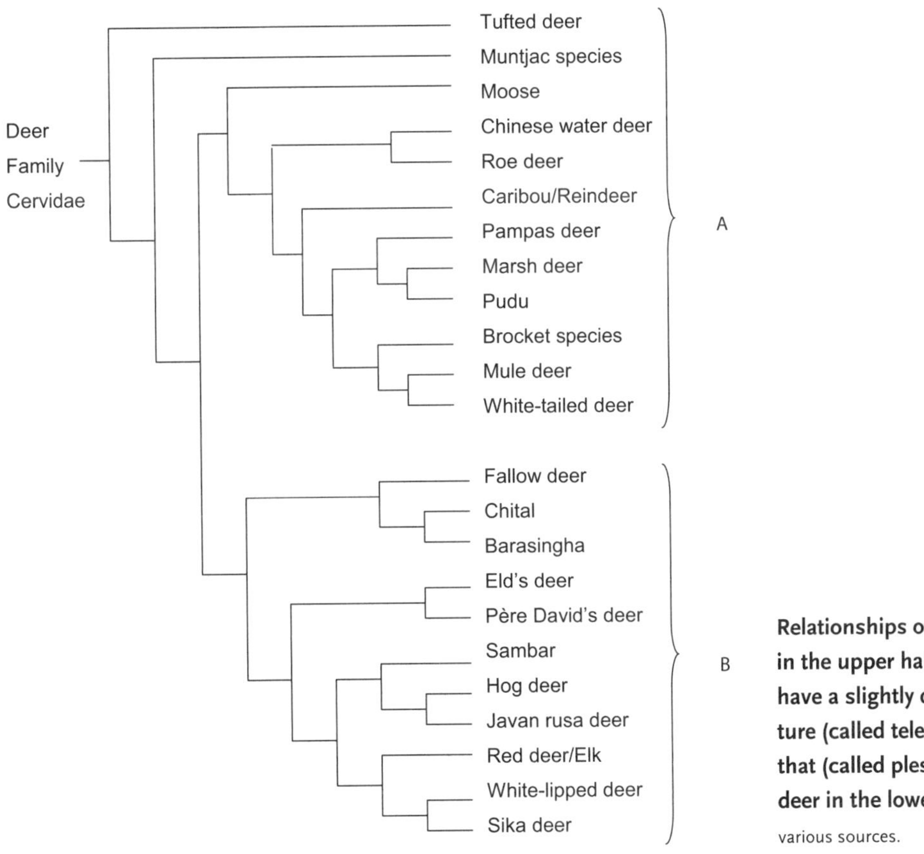

Relationships of living deer. Species in the upper half (A) of the diagram have a slightly different limb structure (called telemetacarpalian) than that (called plesiometacarpalian) of deer in the lower half (B). Adapted from various sources.

sify organisms) now place musk deer in their own family (Moschidae), based on their unique morphological (physical) and genetic characteristics. Results of recent genetic analyses reveal that musk deer are actually more closely related to sheep, goats, cattle, and African antelope than they are to deer. Musk deer differ from true deer in several ways, including having a gallbladder (used to store bile and break up, or emulsify, fat), a single pair of teats instead of two pairs, and males that are somewhat smaller than females. Like Chinese water deer, the seven species of musk deer (genus *Moschus*) lack antlers and have enlarged upper canine teeth. Musk deer also differ from other deer species in that males have a highly developed gland (or pod) that is about the size of a tennis ball and secretes the reddish-brown musk for which they are named. Musk, with its highly pungent scent, has been prized for thousands of years for a variety of purposes, including use in perfumes and traditional Asian medicine. Limited in its supply, but not in the demand for it, musk is, ounce for ounce, one of the most valuable of all animal products. Although synthetic compounds are now available as substitutes, the international demand for musk continues to threaten the survival of some species of musk deer, because of overharvesting and poaching.

What are the most common species of deer?

"Common" can mean that a species has a broad geographic distribution or that it is very abundant regionally—or both. White-tailed deer certainly qualify as one of the most common and successful species. Whitetails enjoy one of the largest natural geographic distributions of any species of deer. They occur from southern Canada, throughout most of North and Central America, and into South America—a greater latitudinal range than any of the others. Local abundances in many parts of the United States have reached pest proportions—although the Columbian white-tailed deer (*Odocoileus virginianus leucurus*) is an endangered subspecies in a portion of its range. Caribou have a circumpolar distribution—in Greenland and in northern portions of the Old and New World—and they have been introduced elsewhere. Unlike whitetails, however, populations of caribou are declining in some areas. Moose also have a circumpolar distribution. Red deer are widely distributed throughout much of the Northern Hemisphere. Fallow deer occur naturally in the Middle East and perhaps in North Africa, but they have also been introduced elsewhere. Free-ranging populations now are found throughout Europe, North and South America, South Africa, Australia, New Zealand, and many Pacific islands.

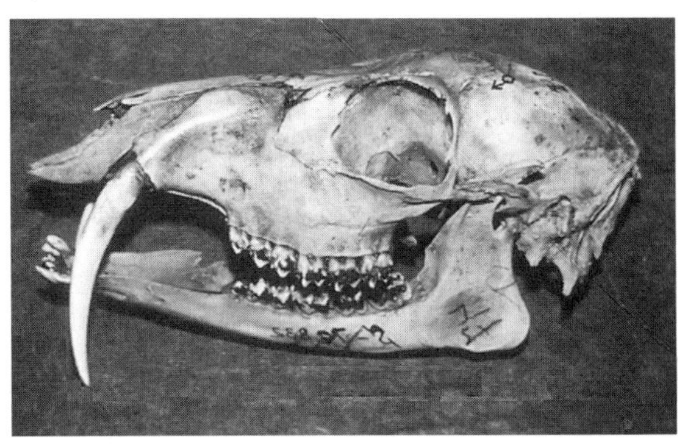

***Left*, Musk deer are very similar in appearance to "true" deer, but they are very different in terms of their anatomy and genetics. *Right*, Skull of a male musk deer.** Drawing by Lisa Russell (*left*) and photo by George Feldhamer (*right*).

White-tailed deer are one of the best known and most popular game mammals in the Western Hemisphere. *Left*, Male (buck). *Right*, Female (doe). Photos by Scott Bauer, U.S. Department of Agriculture (*left*), and Ken Thomas (*right*).

What species of deer are most rare?

Many species of deer are very limited in their abundance and distribution. Several species are endemic—found only in a restricted geographic area. In South America, two species of brocket deer (although not all experts agree that they should be considered as separate species)—the São Paulo brocket deer (*Mazama bororo*) and the dwarf brocket deer (*M. chunyi*)—are endemic. In Asia, several species of muntjac have very restricted distributions, including the Puhoat muntjac (*Muntiacus puhoatensis*) in Vietnam, the leaf deer (*M. putaoensis*) in Myanmar, and Roosevelt's muntjac (*M. rooseveltorum*) in Laos. Likewise, the Bawean deer (*Axis kuhlii*) is limited to one Indonesian island, while the Calamian deer (*A. calamianensis*) and the Visayan spotted deer (*Rusa alfredi*) are endemic to a few Philippine islands. Eld's deer (*Rucervus eldii*) and barasingha (*R. duvaucelii*) are not considered to be endemic, although their numbers are declining significantly. Père David's deer (*Elaphurus davidianus*—called "milu" in China) became extinct in the wild, and the species was restricted only to individuals in zoos and captive herds until the late 1980s, when there were limited reintroductions to Nan Haizi Milu Park near Beijing, and to the Dafeng Reserve on the coast of China's Yellow Sea. The history of Père David's deer is truly remarkable (see chapter 10).

Why are deer important?

Deer are extremely important to humans for sport hunting, subsistence (daily survival), and cultural practices. Many people attempt to keep deer away from agricultural fields, gardens, or expensive suburban landscaping. Crop damage attributed to deer worldwide amounts to billions of dollars annually. In North America and Europe, deer-vehicle accidents also cause billions of dollars in damage, millions of deer are hit, and many people are killed. Deer are a significant food source (prey) for numerous predators. Depending on their population density, because deer are herbivores, they can have markedly positive or negative influences on the ecosystems in which they live.

To varying extents, deer are an important part of the lives of many people around the world. Aesthetically and economically, people have had an integral relationship with deer since the dawn of human history. Paleolithic peoples harvested deer for tens of thousands of years as a source of meat. Their hides provided clothing, footwear, and shelter, and their sinews were used for rope. Antlers were made into tools, decorative objects, or trophies. Deer have a deep aesthetic appeal, as well. Throughout history they have played a significant role in literature, religion, mythology, art, and popular culture (see chapter 11).

As important as deer have been for us across the ages, we probably have had a much greater impact on them. Humans have raised and herded reindeer for thousands of years. In a relatively recent phenomenon, many other deer species are now confined as domestic livestock and raised for meat and antler velvet, for breeding stock, or (in the case of large-antlered males) for hunting. We have also had an impact on deer by moving many species to regions outside of their native range. These free-ranging "exotics" often have caused unanticipated negative consequences for native habitats and animals.

Where do deer live?

Deer naturally occur on all continents except Australia and Antarctica. They extend from 80° N latitude (caribou) south through the tropics to about 50° S latitude for the southern huemul, or guemal (*Hippocamelus bisulcus*). Some deer species are restricted in their distribution—either to the Old World or the New World—but moose, caribou (or reindeer), and red deer (or North American elk) occur in both hemispheres. Along with these three species, white-tailed deer and mule deer (*Odocoileus hemionus*) make up the five species of deer native to North America. There are about 16 species native to Central and South America, three times the number occurring in North America. Five species—moose, fallow deer, red deer, roe deer (genus *Capreolus*), and caribou—occur in Europe (although fallow deer were probably introduced there long ago), while Asia has 28 species. Deer are naturally absent from Australia and southern Africa. Competition with many species of antelope (bovids) south of the Sahara Desert may have kept deer from establishing themselves there. Conversely, the 16 species of deer that occur in Central and South America do so in the absence of many bovids.

Deer have been introduced to Australia, New Zealand (eight different species since the 1850s), and many other islands where they do not occur naturally. Reindeer, for example, were introduced to South Georgia Island in the southern Atlantic (50° S latitude), and to Svalbard in the Arctic Ocean (80° N latitude). Indeed, people have been introducing deer into different parts of the world outside of their native ranges for thousands of years. Today, many species of deer are well established in the wild in North America as exotics (non-native introductions), including Old World fallow deer and sika deer. Sika deer (*sika* is Japanese for "deer") have been an especially successful competitor with white-tailed deer in places where they overlap in Texas, Maryland, and Virginia.

Classification of the deer family (Cervidae)

Order **ARTIODACTYLA**
 Family **Cervidae**
  Subfamily **Capreolinae** (or **Odocoileinae**) (21 species)
   Primarily New World deer; includes species in Europe and Asia
  Subfamily **Cervinae** (28 species)
   Eurasian deer
  Subfamily **Hydropotinae** (1 species)
   Chinese water deer

Note: Alternative taxonomic systems of classification have been proposed.

Just as deer are quite different geographically, the habitats they occupy are also highly variable. Habitats used by a particular deer species are related to their body size and foraging habits, and may include evergreen and deciduous forests, marshes, grasslands, tundra, arid scrublands and deserts, agricultural areas, or rainforests. A few species even thrive in suburban environments. Like all animals, deer occupy only those habitats that provide them with necessary food and water, along with protection both from predators and from adverse weather conditions. Most species have a preferred habitat, but few are restricted to a single habitat type. White-tailed deer are a prime example of a species that occupies a variety of habitats. Even in local areas, a mosaic of different, interspersed types of vegetation is ideal for whitetails, as it is for many other deer species. Nonetheless, many species predominantly inhabit forested areas, often with a dense understory (shrub layer beneath the trees). This is especially true for smaller, solitary species such as pudu, muntjac, brocket deer, roe deer, tufted deer, and sika deer, and even for larger species like red deer. Other grazing species, such as fallow deer and pampas deer (*Ozotoceros bezoarticus*), are generally associated with open grasslands, at least for foraging. Chital (*Axis axis*) and Eld's deer prefer open forests and forest-savanna habitats. Marsh deer (*Blastocerus dichotomus*) obviously prefer marshes, swamps, and other wetlands, as do Chinese water deer, barasingha, hog deer (*Axis porcinus*), and (often) sika deer. Other species, such as moose, may forage in aquatic habitats. The open slopes of mountainous terrain and dry steppes best characterize the habitat of white-lipped deer (*Przewalskium albirostris*) in Tibet and China, and of South American huemuls in the Andes of South America. Mule deer often live in dry, desertlike regions, although different subspecies occupy a variety of habitat types from Alaska to Mexico. Regardless of their preferred foraging habitats, most species of deer are never far from woodlands, which provide cover to escape from predators, as well as shade during the middle of the day.

Broadly defined habitat types are themselves highly variable. For example, the term "deciduous forest" encompasses many local habitat types—including uplands, bottomlands, an open or closed canopy (the uppermost leaf layer of trees), and a dense or open understory—all with different conditions, as far as deer are concerned. The same is true for the many types of wetlands and grasslands. In addition, there are often complex interactions among deer species within a given habitat. In some European forests, for example, red deer, roe deer, and fallow deer may coexist with recently introduced species like sika deer and muntjac.

Each species invariably specializes in their use of the resources around them. They divide up local habitats—often moving to different areas during the day or night as they respond to competition from the other species. For example, this situation applies to elk, black-tailed deer (a subspecies of mule deer), fallow deer, and axis deer at Point Reyes in central California. Deer also modify their use of habitats when the availability of food and cover changes seasonally. Some species migrate from season to season. Caribou are well known for migrations between their summer and winter ranges. Other species—such as mule deer, white-lipped deer, caribou, and huemuls—move from higher elevation, open summer ranges to more covered, relatively snow-free, wooded habitats at lower elevations during the winter. Asian species migrate in response to monsoon rains, moving out of floodplains during rainy seasons and back along rivers during dry seasons.

What is the current classification of deer?

Based on the structure of their limbs, deer are in the taxonomic order Artiodactyla—one of 30 different orders of mammals. There are 10 families in the Artiodactyla; deer are in the family Cervidae. Most specialists recognize three subfamilies of deer: Capreolinae, Cervinae, and Hydropotinae. Deer comprise about 50 of the approximately 5,400 species of mammals worldwide.

The genus and species names taken together (called a binomial, meaning "two names") are always in Latin, and these binomials make up the scientific name of any organism. Scientific names are necessary to avoid confusion, because common names for a species may vary regionally or locally. For example, in North America, "elk" refers to *Cervus elaphus*, but in Europe "elk" refers to *Alces alces*, what North Americans call "moose." Also consider *Przewalskium albirostris*: the common names include white-lipped deer, Thorold's deer, or Przewalski's deer. Red deer have regional common names, including shou, Barbary, hangul, maral, Ruskharian, and Zarkland. Although scientific names may sometimes seem unnecessarily complicated, they often describe morphological or behavioral characteristics of the species. A good example is the scientific name for Chinese water deer, *Hydro-*

potes inermis. *Hydro* is Greek for "water," while *potes* means "a drinker," a clear reference to the preferred wet habitats in which this species occurs. The species name for Chinese water deer, *inermis*, is Latin for "unarmed," and it is certainly appropriate for the only species of deer without antlers. Regardless of their native language, all scientists understand and can communicate with each other using scientific names.

What characterizes the major groups of deer?

Biologists place deer into two major groups, based on the structure of their limbs. Deer walk on the tips of their toenails (hooves), in contrast to bears, opossums, people, and many other species that use the entire surface of the foot when walking or running. Specifically, the third and fourth toes of each deer's foot make up their hooves. Two other toes (called dew claws) occur slightly above the hoof. The second toe is on the inside of the leg, and the fourth toe is on the outside. Also, in deer the third and fourth long bones of the feet (the metapodials) are lengthened and fused to form what is known as a cannon bone. The structure of the vestiges (small remaining bits) of the second and fifth toes is used to group deer into one of two categories. Deer that retain these vestiges nearest the hoof are referred to as telemetacarpalian. In contrast, plesiometacarpalians are those deer that retain remnants of these two toes farther up the foot.

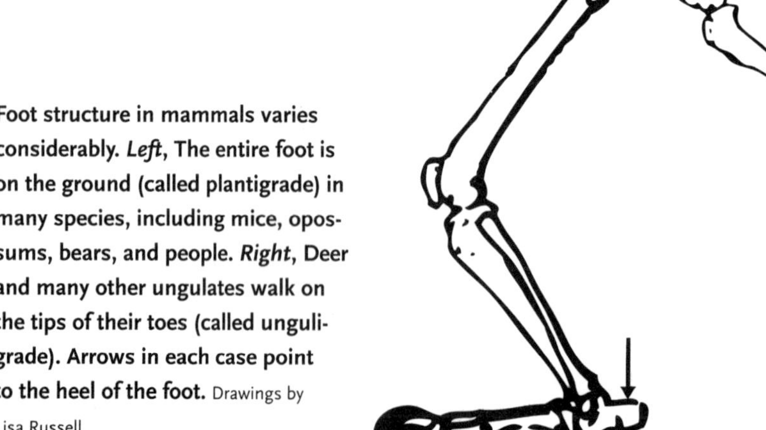

Foot structure in mammals varies considerably. *Left*, The entire foot is on the ground (called plantigrade) in many species, including mice, opossums, bears, and people. *Right*, Deer and many other ungulates walk on the tips of their toes (called unguligrade). Arrows in each case point to the heel of the foot. Drawings by Lisa Russell.

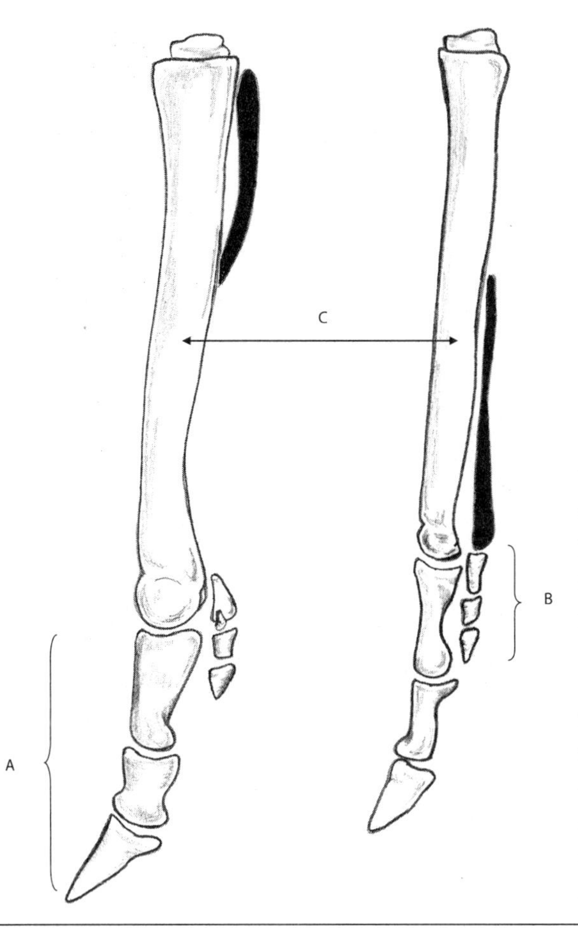

The position of the vestiges of the second and fifth metapodials (in black) generally defines (*left*) plesiometacarpalians (Old World deer), or (*right*) telemetacarpalians (New World deer). (A) shows the three bones (phalanges) in the toe, ending in the hoof; (B) indicates the vestigial dew claw toe; and (C) points to the fused cannon bone. Drawing by Lisa Russell.

When did deer evolve?

Various ancestral lines (lineages) of ungulates (hoofed mammals) with hornlike structures evolved over tens of millions of years. These lineages eventually became recognizable as giraffes and bovids (antelope, sheep, goats), as well as deer. The ancestors of deer arose from primitive ungulates (called condylarths) about 55 million years ago, at the beginning of the Eocene. Fossil remains of the very early ancestors of deer are differentiated from other lineages, based on structures that eventually came to be deciduous antlers, as well as on the size, shape, and chewing surface of the teeth, and the fused bones in portions of their limbs and feet.

Changes in the structure and function of the digestive tract, toward that of a ruminant—for more efficient extraction of nutrition from vegetation—were no doubt occurring as well. Of course, changes in the soft anatomy of the body rarely show up in the fossil record; only teeth and bones usually fossilize. New methods of analysis help to make sense of all the data on morphology, embryology (development from fertilization to a fully-grown fetus), fossils, and genetics, as these early relationships continue to be un-

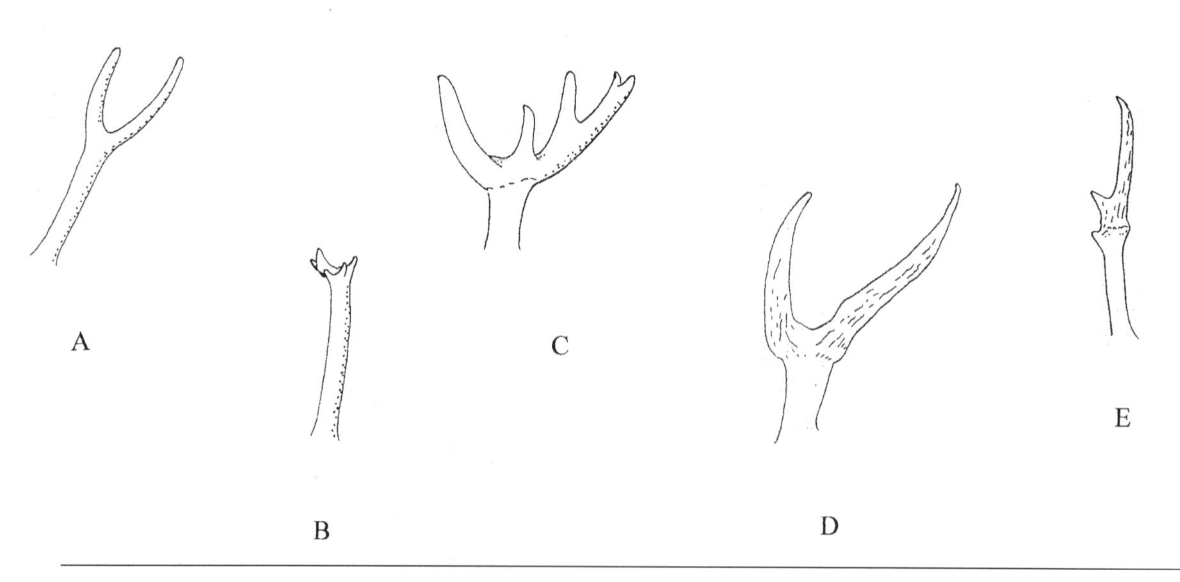

A general sequence of how antlers may have evolved, as shown in five fossil genera of deer: (A) *Procervulus*; (B) *Lagomeryx*; (C) *Stephanocemas*; (D) *Dicrocerus*; and (E) *Euprox*. (See chapter 2, How did antlers evolve?) Drawings adapted from various sources.

raveled. Although he did not write for a general audience, Valerius Geist, in his book *Deer of the World*, discusses deer evolution in great detail.

What is the oldest fossil deer?

The earliest fossils identifiable as deer are about 15 to 20 million years old and date from the early to mid-Miocene. These very early deer are from north-temperate regions of Asia and include an array of genera (the plural of "genus") known only from fossils—such as *Procervulus*, *Lagomeryx*, *Stephanocemas*, *Dicrocerus*, *Heteroprox*, and *Euprox*. They were smaller than most species of deer that are alive today. Deer fossils from North America do not occur until about 4 to 5 million years ago, in the Pliocene. Identifiable genera include the extinct *Bretzia*, as well as *Odocoileus*, the genus of modern mule deer and white-tailed deer. Deer in North America then moved into South America 3 million years ago, after the emergence of the Panamanian Land Bridge connecting Central and South America. These deer later branched out into many different genera and species—more so than are now found in North America. Nonetheless, South American pampas deer and marsh deer look very much like white-tailed deer, and are closely related to them.

Chapter 2

Form and Function

What are the largest and smallest living deer?

Deer species vary considerably in body size. The moose (*Alces alces*) is the largest living deer, with body weights up to 1,760 pounds (800 kg). The smallest deer species is the tiny southern pudu of South America (*Pudu puda*), which is only about 13 inches (34 cm) tall at the shoulder and weighs about 17 pounds (8 kg). As might be expected, the life-history characteristics of a particular species of deer are closely related to its size.

In most species of deer, males tend to be larger. In some species they may be more than 25 percent larger in body mass and dimensions than females. However, exceptions do occur, and the male tufted deer (*Elaphodus cephalophus*) may be slightly smaller than females. Larger male body size in deer (sexual dimorphism) is related to mating patterns and to combat between males over access to females (see chapter 4). Polygynous species (when one male mates with many females) exhibit greater sexual dimorphism, with larger males attracting and breeding with more females than do smaller males. Most large species of deer are polygynous. Body size can also be influenced by geography. Within a given species, individuals at higher latitudes—in regions farther from the equator with colder climates—may have larger body size. This generalization is known as Bergmann's Rule. It is adaptive (useful to an animal), because larger individuals are better able to retain heat than are smaller ones.

The relative body size of New World deer. (A) South American pudu; (B) brocket deer; (C) huemul; (D) pampas deer; (E) marsh deer; (F) tropical white-tailed deer; (G) temperate white-tailed deer; (H) mule deer; (I) caribou; (J) moose.
From V. Geist, *Deer of the World*. Used with permission of Stackpole Books.

What is the metabolism of deer?

Metabolism (chemical processes in the body that sustain life) plays an essential role in the life history of any organism. In deer, as in all mammalian species, there is a direct relationship between body size and metabolism: the larger the animal, the more energy it expends, and the more oxygen it consumes. Metabolic rate often is measured as the amount of oxygen an animal uses in a given amount of time. For example, considering the largest and smallest deer, a pudu that weighs 17 pounds (8 kg) would use about 185 cubic inches (3 liters [3 l]) of oxygen an hour. A 1,760-pound (800-kg) moose would use about 3.35 cubic feet (95 l) of oxygen/hour. What is somewhat surprising, however, is that there is an inverse (negative) relationship between metabolic rates and oxygen consumed per unit of body weight. Even though a moose weighs 100 times more than a pudu, it uses only 32 times more oxygen. Thus, a pudu uses a little over three times more oxygen per ounce of its body weight than does a moose. So although moose consume more oxygen in an hour than a small species such as pudu or tufted deer, the smaller species have higher mass-specific metabolic rates than do larger species. This is called metabolic scaling.

Metabolic rate is also directly related to heart rate—the more an animal exerts itself (when chased by a predator, for example)—the more oxygen it consumes and the faster its heart beats. Smaller deer, because of metabolic scaling, have a faster heart rate than larger deer do. The heart rate for a resting pudu is about 143 beats/minute, while that of a moose is only about 45 beats/minute. In comparison, the heart rate of a healthy human weighing 180 pounds (82 kg) is about 80 beats/minute. Extremes of climate (either heat or cold) place additional demands on metabolic rates. Metabolic rates change seasonally, and behavioral adaptations, such as reduced activity or using sheltered areas, can help reduce metabolic demands during stressful periods.

What is the structure and function of deer teeth?

The structure and arrangement of deer teeth (dentition) reflects their function in cutting and chewing vegetation. Deer are browsers (animals generally consuming the leaves and buds of woody plants) and grazers (those feeding primarily on grasses), as opposed to some other herbivores, like rodents and rabbits, that are essentially gnawers. All deer have a total of either 32 or 34 teeth, including incisors, canines, premolars, and molars. Deer do not have upper incisors. Instead, they have a calloused pad of tissue. All deer have three incisors on each side of their lower jaw (the mandible), although there appear to be four incisors on each side. The tooth furthest from the middle (the lateral tooth) on each side is actually a canine that has become incisorlike in its form and function. Deer cut or tear vegetation using these lower front (anterior) teeth against the pad of the upper jaw (the maxilla). They also use their teeth to strip leaves from tree limbs. Some species have a canine tooth on each side of their upper jaw, such as North American elk (or wapiti) and European red deer (*Cervus elaphus* is the scientific name for both elk and red deer), sika deer (*C. nippon*), huemuls (genus *Hippocamelus*), and others. However, these teeth may not break through the gums in all species. In Chinese water deer (*Hydropotes inermis*), their upper canines do not function in feeding, but instead are used for defense and fighting. Most New World deer and many Old World species do not have upper canines. They were lost through evolution. Occasionally, however, upper canines appear in white-tailed deer (*Odocoileus virginianus*), mule deer (*O. hemionus*), and other species. These evolutionary "throwbacks" are very rare—probably about 1 in 10,000 whitetails, although they may occur more frequently in some areas.

Once deer cut vegetation with their incisors, they crush and grind it with the teeth farther back in the jaw. These molariform teeth (also called cheek teeth) consist of three premolars and three molars on each side of

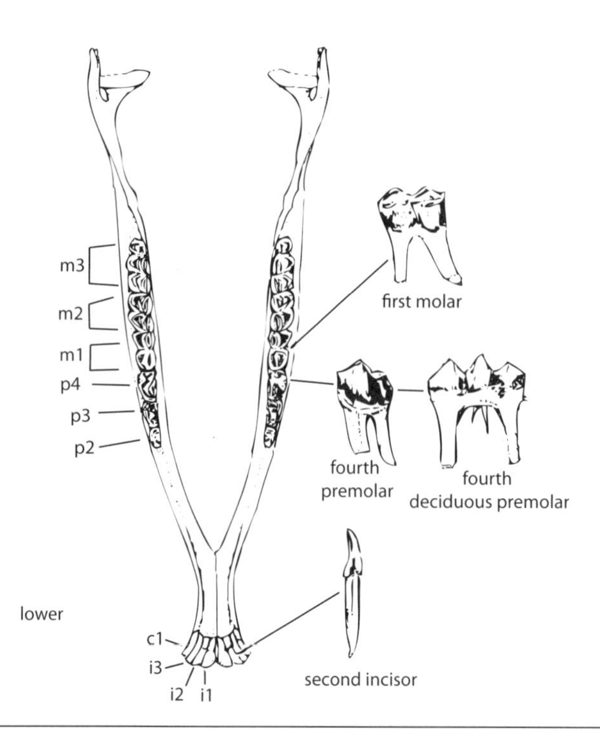

Teeth in the lower jaw of a white-tailed deer. The abbreviations are m = molar, p = premolar, c = canine, and i = incisor; the numbers indicate placement order—first, second, and so forth. Used with permission of the Illinois State Museum.

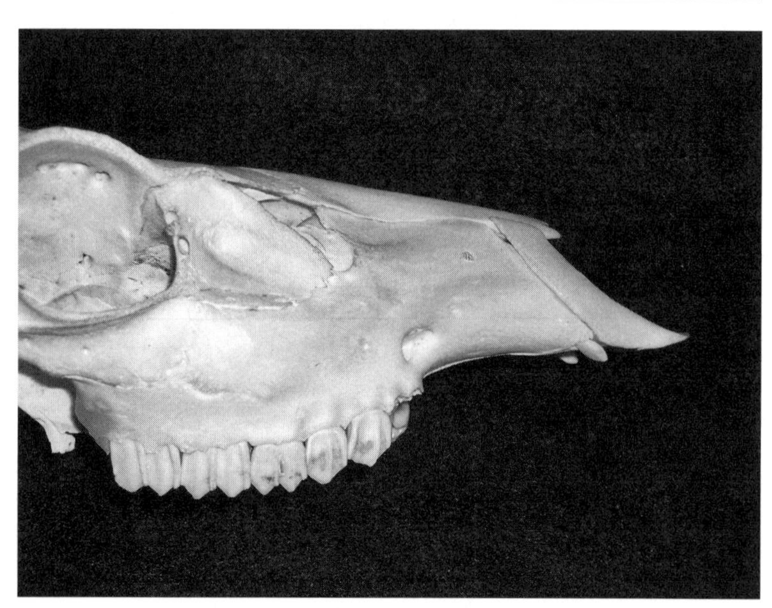

Small upper canine teeth in a sika deer skull. These are generally non-functional. Photo by George Feldhamer.

both the upper and lower jaws. Although structurally they may look very similar, premolars have deciduous counterparts (the equivalent of "baby teeth") that fall out and are replaced by permanent counterparts. Incisors and canines also are deciduous, while molars are not. The gap between the incisors and the cheek teeth is called a diastema. It enhances the ability of an animal to hold and manipulate food in its mouth. A diastema is not unique to deer, as it occurs in many other herbivorous species, including all rodents and rabbits.

The cheek teeth of deer are adapted to their herbivorous diet. Because the vegetation deer feed on is often very coarse, with a high silica content that tends to wear teeth down, they start with high-crowned dentition that extends well above the gum line—a condition called hypsodont. The chewing (or occlusal) surface of the cheek teeth in herbivores is also modified, with specialized folds and ridges in the outer layer of enamel and the inner layer of dentine that give each tooth a crescent shape (called selenodont) that is adapted for crushing and grinding vegetation.

Can deer see color?

Yes, they can, although the blaze-orange color that hunters wear is probably not very obvious to deer. Most of the scientific work on color vision in deer has been done on white-tailed deer and fallow deer, but the anatomy of the eye and the way vision functions suggest that these results most likely relate to other deer as well. At the molecular level, the biology

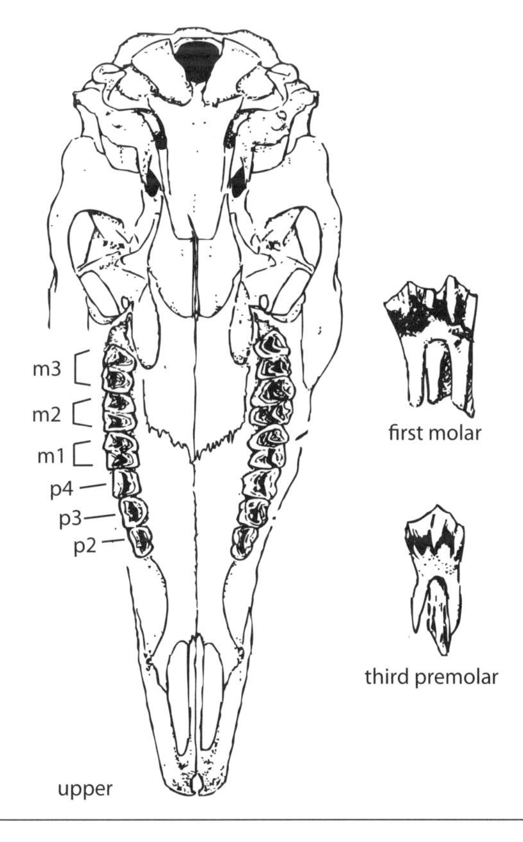

Chewing (occlusal) surfaces of the cheek teeth (premolars and molars) in the upper jaw (viewed from below) of a white-tailed deer. Used with permission of the Illinois State Museum.

of vision in mammals is very complicated, with an intricate network of different types of cells supporting the photoreceptors in the rods and cones in the retina of the eye. The proportion of rods (for black-and-white vision) and cones (for color vision) in a species—as well as the type of cones and their sensitivity to different wavelengths in the light spectrum—determines the extent of color vision. White-tailed deer, as well as fallow deer (genus *Dama*), are dichromatic—they have two classes of cone cells with sensitivity to both short (blue) and medium (yellow-green) wavelengths. Humans are trichromatic, with a sensitivity to longer (red) wavelengths of light that is not known in deer. We also have twice the density of cone cells (almost 13 million per in^2, or about 20,000 per mm^2) in the center of our retinas as deer do. Deer are probably most sensitive to colors that we would describe as blue or blue green, and they may see into the yellow part of the spectrum as well. Most likely color vision is adaptive for deer, because it might complement their keen sense of smell, enabling them to better discriminate among objects in their environment.

Besides perception of color, deer have several other adaptations to enhance how well they see. With an eye on each side of its head, a deer's field of vision is about 310°, even without turning its head, so deer have excellent peripheral coverage. Turning the head only slightly allows a complete view of their surroundings. Like many other nocturnal or crepuscular (active at dawn and dusk) mammals, deer have a *tapetum lucidum* behind the retina. This membrane reflects light back across the retina, increasing brightness and allowing good vision even in conditions where the light is very dim. The *tapetum* is what causes the white "eye shine" of deer at night, such as in the headlights of a car or in a spotlight. Their pupils are large and oval (instead of round), which admits more light. Partly because of the placement of their eyes, deer have fairly poor depth perception. Nonetheless—as any deer hunter knows—deer are famous for their ability to detect even the slightest movement several hundred yards away (especially horizontal movement, rather than objects moving directly toward them). Their senses of smell and hearing, in conjunction with vision, give deer a very acute predator-detection system.

Can deer swim?

Most species of deer are strong swimmers and readily take to water to escape predators, get out of the heat, or forage. Hollow hair shafts, found in the winter coats (pelage) of many northern species, aid deer as they swim by increasing their buoyancy. Caribou (*Rangifer tarandus*) regularly cross rivers during their annual migrations. Sika deer also are very good swimmers. They were introduced to James Island, in the Maryland portion of

Chesapeake Bay, in the early 1900s; the island is about 1 mile (1.6 km) off shore. As their population grew, overcrowding occurred and the habitat deteriorated. Some deer swam to the mainland, where they established a thriving population. Other particularly capable swimmers are moose, sambar (*Rusa unicolor*), barasingha (*Rucervus duvaucelii*), marsh deer (*Blastocerus dichotomus*), Père David's deer (*Elaphurus davidianus*), and, of course, Chinese water deer. These species often are closely associated with water, because they feed on aquatic vegetation or easily swim to other locations.

How fast can deer run?

The running speed of deer depends on their size and gait. Large deer (known as "runners") are faster than small deer (referred to as "hiders"). Hiders seek shelter in dense vegetation rather than attempting to outrun danger. Speed also depends on the incentive an individual has to move fast, such as being chased by a predator or fleeing from another threat (e.g., fire), or when trying to move through dense vegetation. Sambar are large, but they slink away in the forest; they only run when seen in a field. Moose, white-tailed deer, and red deer can run at 45 mph (72 kph) for short distances on level ground. Caribou can attain speeds of 35 mph (56 kph) for a limited time; they can also trot at a more leisurely 8 mph (14 kph) for much of the day. Although they are large, the somewhat ungainly Père David's deer reach a maximum running speed of only about 18 mph (30 kph).

The speed of deer (or any mammal) depends on the length of their stride and the number of strides they take per unit of time. Many species of deer (the runners) have long legs relative to their body size, which allows for an increased length of stride. Also, deer lack a collarbone (clavicle). This, too, helps to lengthen their stride and reduces the shock to the rest of their skeleton when their front legs strike the ground. An equally adaptive physical feature in deer is the astragalus. This anklebone has a pulley-like structure that restricts movement of the limbs to a single plane, an adaptation that again enhances a deer's ability to run.

How high can deer jump?

Jumping ability, like running speed, is related to the size of the deer and the incentive the animal has to get over a barrier. This is an important practical question for people erecting fences in attempts to keep deer away from gardens, landscaping, or airport runways. Fences 6 feet high (2 m) or less will deter small species of deer, but they are insufficient to exclude larger species. White-tailed deer can clear fences 8 feet (2.7 m) high, so what are intended to be "deer-proof" fences along interstate highways are often 9 feet (3 m) high. The taller the fencing, however, the more expensive it is to install and maintain. Often an extension arm, angled out 45°, on the top of a fence is as effective as an additional 2–3 feet of vertical height. Because of the excellent leaping ability of deer, fence designs to exclude them are continually being refined.

What are antlers?

> To the deer hunter, antlers are prized trophies. To the animal lover, they are magnificent ornaments adorning one of the world's most graceful animals. To the zoologist, they are fascinating curiosities that seem to defy the laws of nature. To the deer themselves, they are status symbols in the competition for male supremacy. Antlers are an extravagance of nature. . . . If they had not evolved in the first place they would never have been conceived, even in the wildest fantasies of the most imaginative biologists.
>
> R. J. Goss, *Deer Antlers: Regeneration, Function, and Evolution*

Antlers have intrigued scientists and the public for centuries. The French naturalist Georges Buffon, in *Natural History*—his massive, 44-volume series that he started writing in 1752—stated that antlers were made of wood. Antlers actually are composed of bone, and, like other bones in a deer's body, they are mainly composed of calcium, phosphorus, and other materials, including the protein collagen. Antlers are situated on pedicles (extensions of the frontal bone of the skull). The burr of the antler (the bony rim around the antler base) separates it from the pedicle below, which is living tissue. An antler is no longer living tissue once it is fully grown. Antlers vary considerably in size and shape, both within and among species. Species such as pudu and tufted deer have very short, simple antlers that form a single spike on each side of the head, without any projecting points, or tines. Other species form very large, heavy antlers with numerous tines.

What determines antler size in an individual?

The size of antlers in an individual deer depends on three factors: overall body condition, genetics, and age. Better body condition—which depends on the availability of good habitat with abundant, high-quality food and moderate population density—results in larger antlers. For example, in the eastern United States, the average length and diameter of the antler beam (the main portion of the antler, from the burr to the farthest tip), and the number of points, all increase in yearling white-tailed deer in direct response to increased availability of acorns the previous autumn. North American elk that experience a severe winter as calves have small spike

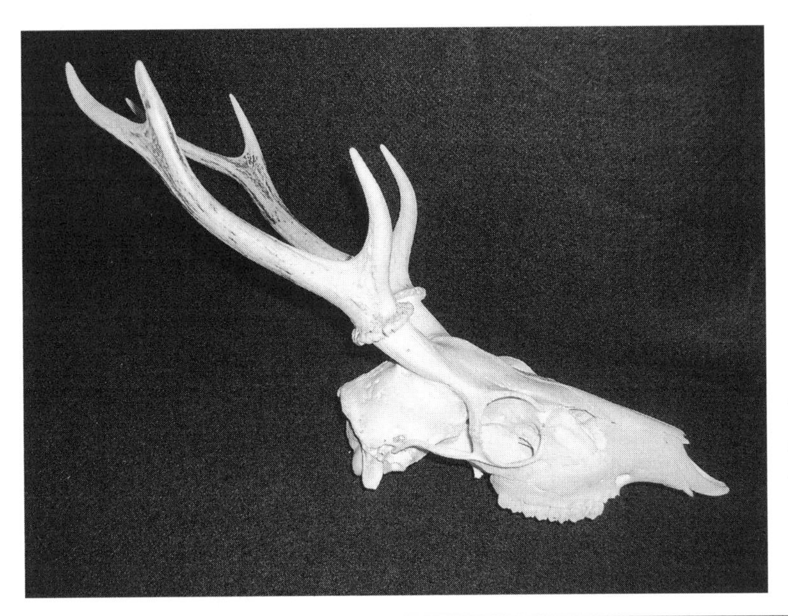

The pedicle on the skull of a sika deer. Pedicles are extensions of the frontal bone and are living tissue. Mature antlers, which are non-living bone, are attached to the pedicles.
Photo by George Feldhamer.

antlers or few points as yearlings, compared with calves having a milder first winter with better resource availability. Antler development also has a genetic component, with some individuals predisposed toward developing large antlers each year. Finally, very young or very old animals tend to have smaller antlers; they are larger in animals in the prime of life. In Alaskan moose, peak antler size is attained by 7- to 11-year-old bulls, and then declines in older animals. Similar trends occur in most other species. It is a common misconception that the number of antler points is equal to the age of the individual. In species with complex antlers, prime-age animals in good condition are much younger than the number of antler points that they have. Antler size is larger in larger species of deer, but this relationship is not directly proportional. In large species, males in the prime of life actually have larger, heavier antlers than would be expected based on their body weight alone, because females may select males to breed with based on the more impressive visual display of increased antler dimensions.

Why do deer have antlers?

The primary function of antlers is their use as weapons and as indicators of status. Males have antlers to allow them to gauge the size, body condition, and social status of prospective rivals that may challenge them for access to estrous females (those that are ready to mate). Antlers allow males to establish dominance hierarchies with other males or to maintain territories so that females will be available to them. Dominance is gained either by intimidating potential opponents through "advertising" one's large

(A) Antlers are short, unbranched spikes in species like pudu, the world's smallest deer. (B) Muntjac may have a very short brow tine (point). (C) Mule deer have distinctive, dichotomously (i.e., paired) branched antlers. (D) The antlers of Père David's deer are unique in the backward direction of their tines. (E) Fallow deer have pronounced flattened (palmate) antlers in older individuals. (F) Caribou have large, extensively branched antlers with a pronounced, asymmetrical brow tine. Drawings by Lisa Russell.

body and antler size, by sparring, or by actually fighting (if necessary). The branched antlers of most species of large deer allow for sparring—tests of strength by engaging antlers (often called racks) with an opponent and pushing or hooking. Actual fighting, and inflicting or receiving injuries, is not the goal. Nonetheless, combatants (often unfamiliar males) can and do get hurt, and mature males often show past battle scars.

Antlers may provide secondary benefits as well. They can function as defensive weapons against predators, although in temperate regions most predation occurs during the winter, after the antlers are shed (cast). They also may help females decide which males represent strong, fit, potential mates. Another possible function of antlers may include acting as radiators. Antler velvet is highly vascularized (contains many small blood vessels), and it provides a large surface area over which to dissipate heat as the antlers grow in the summer. This is not the primary reason for antlers, however, or deer in the tropics would have the largest antlers relative to body size, and they do not.

Do all male deer have antlers?

Chinese water deer are unique among members of the deer family in that the males do not have antlers. In all other species of deer, it is unusual for mature males not to develop antlers. A lack of antlers certainly reduces a male's chances of successfully mating, even though he may be fertile and physiologically capable of breeding. Nonetheless, the rare antlerless, reproductively mature male has been documented in several species, including white-tailed deer, mule deer, caribou, and red deer. Père David's deer have two sets of antlers a year—a large set during the mating season and a smaller set during the non-mating season.

Do female deer ever have antlers?

Antlered females are the norm in caribou, with the more northerly herds having more females with antlers. The antlers of caribou females are cast (fall off) about 10 days after they give birth, while the males cast them following the rut (mating season). Individual caribou cows can be antlerless (a condition known as "genetically bald").

However, an antlered female is a very rare event in most other species of deer, with the possible exception of pudu and some muntjac (genus *Muntiacus*). Antlered females have been documented in the other species present in North America—whitetails, mule deer, elk, and moose—as well as in sika deer and red deer. A recent study of white-tailed deer found about one antlered doe per 3,500 hunter-killed antlered deer, although the ratio varies, depending on the region. Most of these antlered does were in velvet, although a few had polished antlers. Even fewer antlered females occur in other species. Antlered female roe deer (genus *Capreolus*) are much more common, however. The antlers of these females are usually short, unbranched spikes, usually remaining in velvet and generally not being shed. Interestingly, most antlered female roe deer exhibit normal reproductive activity.

Why do female caribou usually have antlers?

The reasons why female caribou have antlers include aspects of geography, habitat, and the behavior of the species. Caribou inhabit higher-latitude regions of the Northern Hemisphere, including the High Arctic (the polar barrens), that have extremely low temperatures and a great deal of snow and ice in the winter. During winter, the availability and quality of food for them is very limited. Harsh winter conditions also make travel difficult and energetically demanding. Most adult female caribou are further stressed during the winter because they are pregnant following the autumn rut (mating season). Antlered females can displace males from the limited available feeding sites because the bulls have cast their antlers earlier, following the rut, while the cows retain their antlers later into the winter. Moreover, pregnant female caribou keep their antlers much later into the season than do non-pregnant females. Interestingly, females that are not pregnant typically do not grow antlers. In the spring, however, antler regrowth starts sooner in males than it does in females, as females don't cast their antlers until after spring calving.

How do antlers differ from horns?

While deer have antlers, horns occur in members of the family Bovidae (African antelope, sheep, bison, goats) and in pronghorn antelope (*Antilocapra americana*). Antlers differ from horns in several ways. Antlers are made of bone, but horns are composed of a sheath made of keratin—a tough, fibrous protein—that overlays a bony core. Antlers are deciduous; species in temperate regions shed their antlers each winter, following the rut, and regrow them in the spring and summer. These growing antlers are covered with haired, highly vascularized skin known as velvet. Unlike antlers, horns are rarely shed. Instead, they grow continuously throughout the life of an animal, with annual growth rings evident on the sheath in many species. A second difference is that horns never branch (those of pronghorn antelope have a short tine, or point, but these animals are in a separate family of their own). Horns may be straight, curved, or spiraled, but they never have the tines seen in most antlers. Finally, in many species horns, unlike antlers, occur regularly in females as well as in males.

What is the yearly cycle of antlers?

The yearly antler cycle of deer living in temperate regions is closely tied to the changing seasons. Although different seasons often mean drastic differences in air temperature, the antler cycle is driven not by temperature change, but by the shifting photoperiod—the relative length of daylight and night. As day lengths increase in the spring, antlers begin to grow. They are cast in the winter (following the autumn mating season), when day lengths become shorter. Most deer from the Northern Hemisphere that are moved to the Southern Hemisphere will eventually shift their antler cycle by six months, as will deer held under artificial lighting conditions that simulate a reversed photoperiod.

Deer living close to the equator may not show this seasonal linkage between (synchrony with) photoperiod and the antler cycle, because there is little annual fluctuation in day-night length. At this latitude, mating may take place at any time throughout the year, there is no well-defined birth pulse, and the herd does not grow antlers in unison. An individual male grows and sheds a set of antlers every 12 months, and the schedule is related to when he was born. Nonetheless, Eld's deer in Southeast Asia have a relatively well-defined, synchronized breeding season between February and May. Other species may show a localized relationship between when their antlers are cast, associated with the heavy rains of monsoons and subsequent droughts.

When and how do antlers grow?

In most species antlers begin to grow soon after the previous set has been shed. The major factor in antler growth during the spring and summer in temperate regions is a complex interaction of testosterone and other hormones in the blood, which is closely tied to photoperiod. Changing photoperiod and declining testosterone levels eventually result in antlers being shed in late winter in most temperate-zone species. Exceptions to the timing of this cycle include Père David's deer, Eld's deer, hog deer (*Axis porcinus*), chital (*A. axis)*, barasingha, and roe deer, all of which grow antlers in the winter and mate in the spring and summer.

In most temperate-zone species, as days lengthen, the antlers begin to grow from the pedicle on top of the skull. The growing antler is living bone and is covered by velvet—highly vascularized skin covered by very short, fine hairs with a velvety texture and appearance on most species. The velvet is a special integument (layer of skin) that has many branches of the trigeminal (facial) nerve. Numerous blood vessels (from the superficial temporal artery) provide the extensive blood flow necessary to supply the large amount of minerals needed for the antlers' extremely rapid growth. Antlers are the fastest growing tissue (including cancer) in vertebrates. The whole system—bone deposition, velvet, nerves, and blood vessels—is very unusual, because the velvet must grow as quickly as the antler does, and, of course, the nerves and blood vessels in the velvet must keep pace as well.

Blood vessels in the velvet may leave their impression in the hard antler after growth is complete, when the velvet dies and is shed. These impressions are especially evident in antlers that are flattened (palmate), such as in moose. Antlers grow for three to five months, although for species like moose and caribou that live at higher latitudes, this time period is shorter. Before antlers ossify (turn to bone) and completely harden, the growing antlers are somewhat spongy and warm to the touch, because of the pronounced blood supply. Once full growth has been achieved for the year, the blood supply to the velvet is first restricted, and then cut off. As the antler ossifies into hard bone, it becomes much denser than it was while growing. The velvet dies and begins to slough off. Deer shedding velvet often vigorously thrash or rub their antlers on branches and vegetation. Thrashing is a dominance display and may continue long after the velvet is shed, however. Animals at this stage of the process are sometimes referred to as being in "tatters." Once the antlers are completely hard and clear of velvet (polished), they no longer have any blood, nerves, or sense of feeling in them, and they are ready for use either as showpieces for display or as weapons for sparring or fighting.

How fast can antlers grow?

Antlers can grow up to 1 inch (2.5 cm) per day in some large species during peak growth. Rapid growth is especially evident in large species at higher latitudes, where the growing seasons (for antlers as well as everything else) are short. At the peak of their development in late June, for example, the antlers of moose in Alaska may gain half an inch (1.3 cm) in length and close to 1 pound (0.4 kg) of mass per day. Metabolic demands are extremely high to generate the calcium, phosphorus, and trace minerals needed for this much growth in such a short time period. If sufficient minerals are not available in the animal's diet, some can be mobilized from skeletal tissue. In smaller deer, with their associated smaller antlers, the drain on their system is not as extreme.

Do antlers of tropical deer differ from those of temperate species?

Antlers on deer from tropical or temperate regions are the same: in their growth, their physical structure and makeup, their uses, and their eventual shedding. Because antler development is closely tied to reproduction and seasonality, their annual growth cycle in deer at temperate latitudes differs from that of deer in the tropics. Seasons in the tropics are based on rainfall, not temperature, and photoperiod changes are much more subtle. So

Clockwise from top, **Antlers begin their growth as small "buttons" in this white-tailed deer. Growing antlers are covered by velvet. Antler growth is complete, and the blood flow to the velvet is cut off. The velvet dies and begins to shed. Cleaned, "polished" antlers.** Photos courtesy of Michael Ondik.

breeding there may occur throughout the year, and tropical deer often have a less clearly differentiated seasonal antler cycle, one generally tied to the monsoon season.

How large and heavy can antlers get?

Moose are the largest living deer, and their antlers can span 80 inches (203 cm) and weigh 77 pounds (35 kg). Antler-beam length in North American elk (wapiti) can reach 64 inches (163 cm), and their antlers can weigh 34 pounds (15.5 kg). Antler size and weight in living deer certainly remain "monuments to calcium metabolism," although this is not nearly as impressive as the antler size reached by some Pleistocene species (see "How did antlers evolve?").

How are antlers shed?

Mechanically, antlers drop off because a shear plane forms between the base of the antler and the pedicle on which it is attached. The shear plane is created through the action of specialized cells (called osteoclasts) in the end of the pedicle that resorb (demineralize, or dissolve and take in) bone. This directly coincides with the amount of testosterone in the blood, which is greatly reduced following the rut. Antler casting is a fairly sudden phenomenon. The connection between antlers and their pedicles remains very strong until just days before they fall off. Larger, more mature, dominant mule deer and whitetail bucks, bull moose, and red deer stags begin their antler growth earlier than younger males, and, depending on habitat and demographic conditions, they may cast them sooner as well. Age is not the only factor in the timing of when antlers drop off. It also depends on declining body condition (because of the stress of the rut and associated poor nutrition), reproductive status, the length of time estrous females are available, and other factors. As a result, antler shedding in a given herd may extend over several months.

Why shed antlers and regrow them every year?

There are several factors that make it useful (adaptive) for deer to cast and regrow antlers each year. Nonetheless, given the substantial amount of energy invested by a male in antler growth, it seems to be a terrible waste to shed them after the breeding season, only to start the entire growth process again within a few months. Because antlers are used to fight or spar with other males to establish dominance for mating and to attract females, they are especially critical to males. Fighting may result in antler breakage, so if

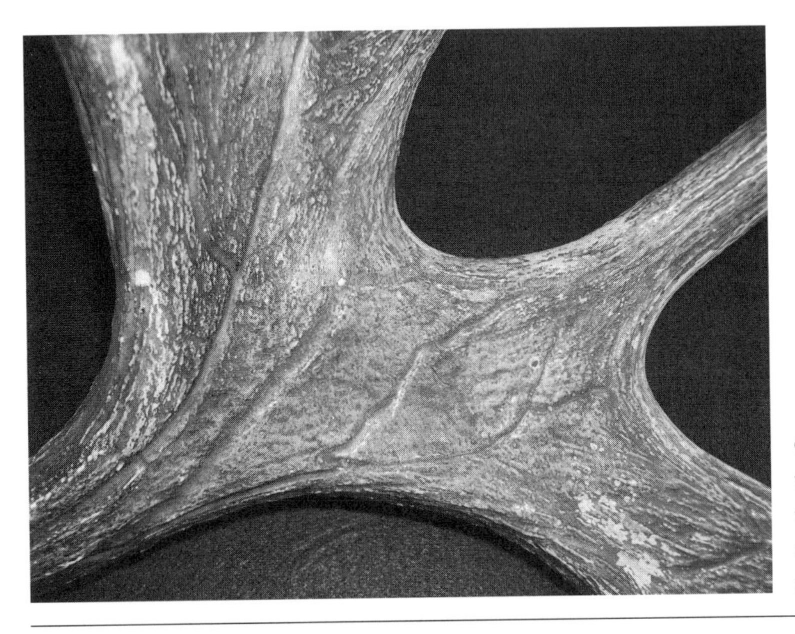

Canal-like impressions, showing where the blood vessels were in the velvet of the growing antler, are easily seen in this section of moose antler. Photo by George Feldhamer.

they were not replaced yearly, a broken antler (or cumulative breakage over several years) could seriously reduce the chance of mating for an otherwise healthy male. Also, in species that inhabit northern or temperate regions, winter—which follows the rut—is a particularly stressful time. The added energy needed to carry antlers through the winter, when males are already metabolically stressed, is not adaptive, because antlers can be heavy (with larger necks needed to carry them), and they are no longer needed for mating. Casting antlers may allow males to be less visible to predators in the winter, although predation on males is often greater than females during this time of year, because males are already in poor condition following the stress of the mating season. Also, as a male increases in body size each year, his antlers can get larger only if they are cast and grown again annually, which accurately reflects his current body condition. Males with large antlers—which can serve as a signal of a male's nutritional condition and suitability as a mate—may breed more often than males with smaller antlers. So, despite appearing to be an energetically wasteful practice, there are good reasons why the yearly cycle of shedding and regrowing antlers is adaptive in deer.

Why are relatively few shed antlers found on the ground?

Given the high numbers of deer in some areas and the fact that males shed their antlers annually, you might expect to trip over piles of antlers every time you walk through the woods. This is not the case, because hunters often harvest males before they shed their antlers. Also, many species chew on cast antlers for the calcium and other minerals they contain, especially when these animals inhabit places where the soils are deficient in minerals. Antlers are a favorite of rodents, rabbits, and other smaller animals. Often, when cast-off antlers are found on the ground, it is evident they have been gnawed on. Nonetheless, people often do find shed antlers. For some, looking for cast antlers is an enjoyable hobby, and the North American Shed Hunters Club, established in 1991, keeps a record book of the largest shed antlers that have been recovered.

Are antlers always symmetrical?

The size, shape, and number of points on the left and right antlers of an individual rarely are perfectly symmetrical. Yearly differences in the beam length of an individual's antlers, and the number and length of their points, probably occur in all species with complex antlers; small deer with simple antlers exhibit greater symmetry. Such antler differences have been documented in many species, including moose, sika deer, fallow deer, and mule deer. The brow tines of caribou are well known for their asymmetry, with the left brow tine (the shovel) usually being much larger than the right. Both tines are the same size in only about 15 percent of caribou. Annual differences are evident in the shovels of individual animals, as each grows a new set of antlers every year. There is a slight asymmetry in moose antlers, with more tines usually occurring on the left antler, as well as differences in the length and width of the palm (the spade-shaped portion). The extent of asymmetry may increase with age.

How do antler anomalies occur?

Anomalies are pronounced irregularities or gross malformations that result from an injury during the developmental stage of antler growth, from an injury to the animal's body, or from an underlying genetic cause. Antler anomalies that are genetically based would be expected to occur in the same individual in successive years, and possibly in future generations. However, serious malformations that negatively impact the ability of an individual to mate would probably not remain in the gene pool.

Antler anomalies more commonly result from a direct injury to the growing antler while it is in velvet. Such abnormalities usually are present on one side only (unilateral) and are not seen in successive years. Abnormalities sometimes occur on one or both antlers. In white-tailed deer, they result in bulges on the main antler beam, called "acorn antlers," which are highly prized by hunters. Reports of trauma-induced antler anomalies are

Signs of gnawing by rodents are obvious on this piece of antler from a white-tailed deer. Photo by George Feldhamer.

fairly common throughout the scientific literature, and they can occur in all deer species.

An injury elsewhere to the body can also result in an antler anomaly. Malformed antlers on one side have occurred following an injury somewhere on the opposite side of the body. Such contralateral anomalies have been seen in numerous species, including red deer, sambar, and white-tailed deer. The reasons for this contralateral effect remain obscure. Antler abnormalities may also result from injury to the pedicle, either from fighting or from infections. One final cause researchers have suggested for antler anomalies is that they form as a result of abnormal hormone levels; anomalies have been linked to abnormal testicular development in red deer, mule deer, and other species.

How did antlers evolve?

Beginning over 20 million years ago, in the early Miocene, many fossil ungulates (hoofed mammals) had hornlike appendages. These outgrowths were small and without tines, and possibly they were used just to lunge at or ram an opponent. Over time, these knobby head ornaments diversified, taking different forms in different groups of ungulates, including antlers in deer. Antlers probably began as simple spikes that were not shed. Through evolutionary time, antlers became increasingly more specialized and elaborate in their length, the number of their tines, and their palmation. The early form of antlers (protoantlers) in prehistoric *Procervulus*, from about 18 million years ago (MYA), were forked. They lacked a burr, however,

The massive antlers of the giant Irish elk swept out and back from the head—much more useful as a display to females than for fighting with other males. From W. B. Scott, Proceedings of the Academy of Natural Sciences.

so it remains unclear whether they were shed like true antlers. Early antlers in *Lagomeryx* (from 17 MYA) had a long spike with numerous, very short prongs on the end (see figure on p. 14). An array of longer terminal points and a shorter pedicle were evident in *Stephanocemas*, from 15 MYA. At about the same time, antlers in *Dicrocerus* showed a reduction in terminal points and two longer, heavier tines. *Dicrocerus* shed its antlers, but they were structurally different from the antlers in deer today. *Euprox furcatus* had two-pronged antlers, with a trend toward having a main beam, closer to the antler structure of modern deer. These early genera (plural of "genus") of deer all had a small body size; small, simple, widely spaced antlers; and enlarged upper canines. All these features are retained today in the relatively primitive muntjac and tufted deer. Later Miocene fossil genera, such as *Ligeromeryx*, had greater branching and structural variation in their antlers, a clear adaptation to engage opponents in sparring while protecting the owner's head. The trend in late Miocene deer continued toward shorter pedicles and longer antlers when these deer radiated from Asia into Europe and North America.

The Pleistocene (starting about 1.8 MYA) was a time when a number of large forms of deer developed worldwide, with correspondingly huge antlers. For example, the giant "stag-moose" (*Alces* [*Cervalces*] *scotti*) was somewhat larger than moose today, but with more complex, palmate antlers 5 feet (1.7 m) across. In the Old World, the giant Irish elk (*Megaloceros*) boasted the largest antlers of any deer that has ever existed. They extended 12 feet (3.6 m) across and weighed as much as 99 pounds (45 kg)—10 times the weight of the skull that carried them. Antlers of this enormous size and shape may have been used more for display than for active combat. The range of *Megaloceros* extended far beyond Ireland, and the species may have persisted in a few areas until about 3,000 years ago.

Chapter 3

Deer Coat Colors

What are the functions of the coat in deer?

The coat (or pelage) of a deer serves a variety of functions, including acting as a barrier to biting insects and protecting against the harmful effects of solar radiation. Primarily, however, the function of the pelage is insulation. Insulation is critical for all but tropical deer species during the winter, especially for those at higher latitudes, such as caribou (*Rangifer tarandus*), red deer (*Cervus elaphus*), moose (*Alces alces*), and southern huemul (*Hippocamelus bisulcus*); or those at high elevations, such as southern pudu (*Pudu puda*) and white-lipped deer (*Przewalskium albirostris*). The long, coarse outer coat of guard hairs and the thick, wooly underfur of these species allow them to endure very cold temperatures. For example, white-lipped deer can withstand the freezing conditions that occur at elevations of 16,200 feet (5,400 m) or more in the mountains of Tibet, where they experience very few frost-free days annually. The length and density of the hairs and the insulating capacity of the coat vary seasonally in most species (see "Seasonal Differences"). A deer's coat also functions as camouflage. The concealing (cryptic) color and spotting patterns that occur in most young deer help them hide from predators. Anyone who has ever searched for newborn fawns knows how effective this coat pattern is—they are exceptionally difficult to locate visually.

Third, pelage can communicate warnings among deer in a group and alert them to potential danger. This is certainly the case with the well-known tail flagging behavior of white-tailed deer (*Odocoileus virginianus*) when they are alarmed and flee (see chapter 4)—a behavior also seen in South American marsh deer (*Blastocerus dichotomus*) and pampas deer (*Ozo-*

***Left*, Unlike red deer, caribou have a pronounced white neck mane. *Right*, Their thick winter pelage (coat), including the heavy neck mane, allows red deer to make it through the winter.** Photos courtesy of Karen Laubenstein, U.S. Fish and Wildlife Service (USFWS) (*left*) and by Rory Putman (*right*).

The cryptic (camouflaged) spotting pattern of white-tailed deer fawns can make them difficult to see in heavy underbrush. Photo from Forest-Wander Nature Photography.

toceros bezoarticus). The white or cream-colored rump patches of several species, including red deer and elk (*Cervus elaphus*), sika deer (*C. nippon*), and roe deer (genus *Capreolus*), serve the same function. When an individual becomes alarmed or senses danger, it flares its rump patch as a warning signal to others in the herd. This flaring of hairs on the rump, a movement that looks like a flower petal suddenly opening, results from the contraction of the small muscles at the base of each hair (the *erector pili*) in the dermis (the layer of tissue beneath the outer layer of skin). A flagging tail or flaring rump patch is absent in smaller, solitary deer species that inhabit dense vegetation. Finally, the coat provides buoyancy when deer swim, because the center of each hair shaft is hollow and filled with air. Many species—Chinese water deer (*Hydropotes inermis*), caribou, marsh deer, and others—are closely associated with water, and all deer are good swimmers.

The white rump patch is evident in several species, including these elk.
Photo by Jim Leupold, USFWS.

The structure of hair (or fur—the terms are synonymous) is the same in deer as it is in most other mammals. A typical hair has three well-defined layers. The center, or core, of a hair shaft is called the medulla. It has a variety of different structural patterns in different species, but it is usually hollow in deer. The cortex makes up most of the hair shaft, and it is formed from tightly packed cells that surround the medulla. Finally, a transparent, thin, scalelike outer layer makes up the cuticle of each hair. Deer have three types of hair. Guard hairs form an outer layer of long, coarse body hairs. The guard hairs of deer are called awns; they have a weak base and a heavier tip, so they lay down in one direction. Guard hairs overlay the shorter, finer, more closely spaced underfur that makes up most of the pelage of deer. In addition, like other mammals, deer have long, stiff, specialized hairs (vibrissae) around the snout (the whiskers) and eyes (the eyelashes). These hairs are connected to nerves and serve a tactile function—that is, they convey a sense of touch.

What causes the different coat colors of deer?

The coat color of a deer is the result of genetically controlled physiological and biochemical processes. The distribution and relative density of different types of pigment granules (melanins) give hair its color. Melanins are contained in the cortex of a hair shaft. Darker black and brown pigments are produced from eumelanins, and the lighter red and orange colors come from pigments called pheomelanins. Different proportions of each pigment throughout the pelage produce the coloration seen in each species of deer. Banded patterns (called agouti) result from an alternating distribution of these pigments. A lack of pigment results in a white coloration, while an excess of melanins causes normally brown deer to be black. Melanistic (dark) individuals occur in numerous species of mammals, as do animals that are all white (albinistic). White individuals, however, are not necessarily albinos. Ultimately, the coat colors we see in every species of deer result from external, environmental selective pressures and evolutionary responses unique to each species.

How are hair colors determined genetically?

Cells called melanocytes contain pigments that are responsible for melanogenesis—that is, the development and production of the color patterns seen in the skin and hair of deer and other animals. During embryological development (the growth of the fetus), melanocytes eventually migrate to the follicles in the skin that produce the growing hairs. Melanogenesis results from an extremely complex network of biochemical interactions that involve numerous agents (regulator enzymes, receptors, coenzymes, and other proteins). Research on these pathways has primarily involved work on humans (focused on malignant melanomas, i.e., cancers) and mice. There are several enzymatic regulators and at least 150 genes that affect coat color in mice. The major regulatory proteins, however, are essentially the same in mammals, so the processes in deer probably are similar to those in mice. When abnormal coat colors (such as albinism or melanism) occur, they are the result of an interruption (error) in some part of the cascade of biochemical reactions that would occur under normal circumstances. Melanogenesis is a critical aspect in how pelage colors and patterns in individual deer ultimately function for camouflage, social communication, and protection against solar radiation.

What about patterns of coat color?

In most species of deer, the general coloration of the adult coat on the upper (dorsal) body is reddish brown, while the lower (ventral) body color is a lighter cream or white. This countershading pattern may serve as a form of camouflage. When viewed from below, deer tend to appear light and blend in with the sky; when viewed from above, they look darker, like the ground. Countershading occurs in many small mammals, such as rodents, and it probably functions more as an antipredator adaptation in these species than it does in deer. Nonetheless, shading and disruption of how predators see deer are probably important factors in coat coloration. Several species of deer—including North American elk, mule deer (*Odocoileus hemionus*), moose, and others—have a very dark belly and no contrast between the color of their upper and lower pelage. The rump and tail coloration in some species is distinctive and may be used to communicate with other members of the herd. The adult coats of a few species of deer—specifically fallow deer (*Dama dama*), chital (*Axis axis*), sika deer, and, to a lesser extent, barasingha (*Rucervus duvaucelii*)—retain the heavy spotting found in newborn animals, and this spotting in adults is most apparent during the summer.

Besides basic coat-color patterns, some species have other, quite distinctive pelage characteristics that are sometimes associated with their common names. Tufted deer (*Elaphodus cephalophus*) have a pronounced mane, or tuft, of bristlelike hairs on their heads (the species name, *cephalophus*, means "head crest"). Likewise, white-lipped deer have a distinctive band of white hair around the front of their muzzle (the species name, *albirostris*, means "white nose"). They are also unusual in having a clump of hair between their shoulders (the withers) that points in the opposite direction from the rest of the coat and looks like a slight hump. In contrast to white-lipped deer, marsh deer have a distinctive black band across the nose, as well as black "stockings" on each leg. One of the more unusual pelage characteristics occurs in moose. Their familiar dewlap, or bell, is a long, hair-covered flap of skin. It is longer in males than in females, and also more rounded, because it often freezes. This bell may be as long as 20 inches (51 cm) or more in males, although it apparently becomes shorter as an individual ages.

Are there age-related differences in coat color?

Newborn deer of almost all species have a spotted coat, while their parents usually do not. A spotted pelage pattern helps the fawn or calf remain camouflaged on a sun-dappled forest floor or in a brushy understory (lower-level vegetation growing below the trees) until they are old enough to outrun potential danger. There are a few exceptions, however. Newborn marsh deer, huemuls (genus *Hippocamelus*), caribou, and moose have solid-colored coats. This non-spotted pattern is particularly unusual in marsh deer, because their young are hiders rather than runners. Interestingly, newborn northern pudu (*Pudu mephistophiles*) have spotted coats, as would

The large dewlap (or "bell") below the chin is evident on a bull moose from Alaska. Also note the characteristic palmate (flattened) antlers. Photo by Donna Dewhurst, USFWS.

Unlike most young deer, moose calves do not have a spotted coat.
Photo courtesy of Catherine Putman.

be expected, but newborn southern pudu (*P. puda*) have solid coats. After they shed their pelage for the first time, young deer of most species lose their spots and take on the solid coat-color pattern of the adults.

Do coat colors change in different seasons?

There are definite seasonal differences in the color and characteristics of deer coats. Deer generally replace (molt) their pelage twice a year, as do most other species of mammals. Like other seasonal phenomena, molting is controlled by hormones, and it is closely tied to changes in photoperiod (the relative length of day and night) and temperature in non-tropical environments. Molting is a highly functional adaptation, because the hairs break and wear down as the pelage is exposed throughout the year to rain, wind, and the bleaching effect of the sun. Other than in tropical regions, deer may face substantial differences in seasonal temperatures. The spring molt results in pelage that is often brighter in color, with shorter, finer, more widely spaced hairs. This coat is less insulating, a useful feature for the upcoming summer. In many species their pelage may appear ragged and moth-eaten during spring molt. The autumn molt results in a heavier, thicker, usually coarser coat, accompanied by a shaggy neck mane in species such as red deer, sika deer, barasingha, and elk. Winter coats—often darker and grayer in color, at least in temperate regions—are more insulating than summer coats. In deer species where the adults are spotted, the spotting of their winter coats is muted and much less obvious than it is in the summer.

Some tropical species may only molt once annually. Although caribou are certainly not tropical in their distribution, they also molt just once a year, in the spring, losing their winter coats and growing sleeker summer pelage. The hair in caribou lengthens in autumn: white-tipped guard hairs create their winter pelage, and a heavy white neck mane forms in bulls.

Is there geographic variation in coat color within a given species?

Within every deer population there are differences in their typical coat color, but certain species can be exceptionally variable in the color of their pelage, including geographical differences among subspecies. For example, woodland caribou (*Rangifer tarandus caribou*) have a particularly dark pelage, while Peary caribou (*R. t. pearyi*), living in the polar barrens of the High Arctic, have very light coats. Mule deer inhabiting deserts have pale coats, but those in forests have a darker pelage. Probably no species has a greater range of coat colors than fallow deer, however. Four varieties of coat color occur, often in the same herd. The most typical color is a rich brown on the upper body, with hundreds of white spots, and a lighter belly and legs. The common name "fallow" probably came from the Old English word *fealou*, meaning "brownish" or "reddish yellow." A paler variation of the typical brown coloration of this species is called menil. Fallow deer can also have a white pelage. These individuals are not albinos, however, as they have dark (rather than pink) eyes. Finally, there are black (melanistic) fallow deer, with dark brown spots—visible at close range—on their backs. Many fallow deer, whether in the wild or in parks, exhibit gradations between these four basic coat types.

Like fallow deer, other species can have atypical white or black coats, although this is usually very rare. Sika deer on the Ryukyu Islands of Japan are highly melanistic. The typical black pelage of this subspecies results from their relative geographic and genetic isolation. Melanistic deer may be prone to overheating in the summer, especially when they are in more tropical latitudes. Conversely, some of the white-tailed deer in Seneca County, New York, are well known for being all white, even though they are not albinos. Likewise, individual white roe deer, red deer, and moose may be fairly common in restricted geographic areas.

True albinos have been documented for many species of deer, but they are usually very rare. Albinism results from a genetically recessive (nondominant) condition. Without any color pigments (melanins) in them, the

This white-tailed deer from Patuxent, Maryland, has a piebald pelage pattern. Photo from the USFWS.

eyes, nose, antler velvet, and any other tissue where blood vessels are close to the surface appear pink to an observer. True albinos are rare, in large part because they have poorer eyesight than normal deer, as well as because their pure-white color is much more visible to potential predators.

Piebald deer exhibit another variation in coat color. Their pelage is white, with asymmetrical blotches of brown fur spaced irregularly over the body. Like albinos, this piebald condition results from a genetic anomaly, but here pigment occurs in the eyes and nose, so these features are not pink, as in albinos. Piebald deer often have a number of associated physical deformities, including shorter legs, a crooked spine, a shortened lower jaw (parrot jaw), a bowed or humped nose, and irregular internal organs. Piebald syndrome seems to occur equally among males and females. It is more common than albinism or melanism, but it is still rare—less than 0.2 percent of hunter-harvested white-tailed deer in the eastern United States are piebald. Piebald mule deer and roe deer also have been documented, and this condition probably occurs as well in several other species of deer. Piebald reindeer lack deformities, and their coats are highly prized by reindeer herders. Anomalous animals may be more common in protected populations, because their numbers are not reduced by hunters.

Chapter 4

Deer Behavior

Are deer social?

All deer are social, because they rely on males and females coming together to mate and on maternal care for their offspring. However, species of deer differ in the degree of social cohesion (togetherness) among adults. For deer that are territorial, a male and a female may share a home range. This behavior is common in smaller species, such as brocket deer (genus *Mazama*) and southern pudu (*Pudu puda*) in South America, tufted deer (*Elaphodus cephalophus*) and muntjac (genus *Muntiacus*) in Asia, or roe deer (*Capreolus capreolus*) in Europe. The Chinese water deer (*Hydropotes inermis*) is probably the least social deer species; territoriality is observed only in males during the mating season, and there are no extended pair bonds between individuals. For the larger species, gregariousness and the degree of their social behavior depends on the environment. Species in more open habitat, such as chital (*Axis axis*), caribou (*Rangifer tarandus*), and fallow deer (*Dama dama*), tend to have larger social groups. Deer in more dense forest, such as moose (*Alces alces*), white-tailed deer (*Odocoileus virginianus*), and sambar (*Rusa unicolor*), have a basic social unit of mother and fawn, with some loose association of related females.

For maternal care, all adult females respond to calls from their fawn or calf and provide grooming and nourishment. For species that occupy more open habitats, newborns must be precocial—able to develop rapidly and move to follow the mother and keep up with the herd. Precocial young are most pronounced in caribou, where calves can struggle to their feet and follow their mother within hours after their birth. For forest species, and for smaller deer, a mother will leave her young and return at periodic

Some species, including these fallow deer, form large social groups. Photo courtesy of Jackie Pringle.

intervals for feeding. As soon as the young animals have matured enough to follow their mothers, they begin learning to identify edible food, foraging locations, and the trail system within their home range. For deer that form large herds, like chital and caribou, maturing animals remain within the herd and have no recognizable bonds with their mothers after they are weaned. For most species not found in large herds, once the young reach maturity they will move away from their mother's home range.

Dispersal may not be the same for both genders, as young females may remain in close proximity to their mother's home range, producing informal matriarchies. In white-tailed deer, for example, small groups of does observed throughout most of the year are usually related individuals derived from a single matriarch.

Male yearlings tend to disperse from the birth area and either live alone or in small groups apart from the maternal herd. The bachelor herds of red deer (*Cervus elaphus*) are a good example of this phenomenon. For deer species that have been studied, there appears to be no particular bond between fathers and their offspring. Some paternal care, such as defense against predators, may occur in the smaller, more territorial species, but researchers have not yet observed such behaviors. In larger herds, males may defend the group from predators, but this defense is not limited to specific offspring.

During the mating season, social behaviors among deer have three main roles: signaling that females are ready to mate, moderating aggression between males and females attempting to mate, and regulating competition among males for access to females. Deer are famous for epic battles between sparring males during the time leading up to the mating season (see "Do deer fight?"). Sparring is especially common for the larger deer species, where there is more sexual dimorphism—that is, a greater difference in body size between males and females. Female red deer and Eld's deer (*Rucervus eldii*) appear to select males based on antler size. For all species, females are the ones that ultimately select the males they end up mating with, for one simple reason: no matter how much larger than the female, a male cannot mate with a doe that will not stand still.

Most social behavior among adults during the breeding season takes place between males competing for females. Females signal their receptivity through pheromones (chemicals that trigger a social response) deposited in gland secretions, and by how they behave toward males. Male courtship behavior may include a ritual approach to a receptive female, such as the neck-low, head-raised posture seen in most New World deer and in red deer. Male interactions usually include a head-turned display of antlers, a ritualistic approach, and the locking of antlers. The males of some species, such as mule deer (*Odocoileus hemionus*), signal their virility by urinating on their hind legs; others, like moose, red deer, and Père David's deer (*Elaphurus davidianus*), urinate on their necks. Both males and females can detect pheromones by means of an accessory olfactory (scent) organ, called the vomeronasal organ. By throwing back the head and curling the upper lip, fluids and aerosols (fine droplets suspended in the air) that are deposited in the mouth can reach the vomeronasal organ through canals located on the floor of the nasal cavity. This behavior is called flehman. Flehman is most often observed during the mating season, where it is performed by male deer after contact with a female's urine. For herding species, like fallow deer and red deer, the males herd the females before and during the mating season, and they employ the same displays of dominance toward females that they use for competing males.

Some social behavior is related to defending against predators. When sighting a predator or when becoming alarmed, several species have an alarm call or display that alerts other herd members to the danger. When fleeing a predator, for example, a white-tailed deer will snort loudly, raise its tail, and spread the hairs on its white rump patch (see "flagging" discussioin on page 52). Similarly, sika deer (*Cervus nippon*) and red deer or elk (*Cervus elaphus*) flair their white rump patches when alarmed. It is unclear

A red muntjac (*Muntiacus muntjak*). Muntjac are also called barking deer.
Photo courtesy of M. Monirul H. Khan.

Indian hog deer (*Axis porcinus*). Photo courtesy of Randy Rieches.

Southern huemul, or guemal (*Hippocamelus bisulcus*). Photo courtesy of JoAnne Smith and Werner Flueck.

Marsh deer (*Blastocerus dichotomus*). Photo courtesy of Arnaud Desbiez, Royal Zoological Society of Scotland.

Pampas deer (*Ozotoceros bezoarticus*). Photo courtesy of Mario Santos Beade.

Reeve's muntjac (*Muntiacus reevesi*). Photo courtesy of Li Sheng.

Northern huemul, or taruca (*Hippocamelus antisensis*). Photo courtesy of Ramiro Rodriguez.

Tufted deer (*Elaphodus cephalophus*). Photo courtesy of Li Sheng.

White-tailed deer (*Odocoileus virginianus*). Photo courtesy of Li Sheng.

Moose (*Alces alces*). Photo by Donna Dewhurst.

Elk (*Cervus elaphus*). Photo by Kaldari.

Female sitka black-tailed deer (*Odocoileus hemionus columbianus*), a subspecies of mule deer in Alaska. Photo by Steve Hillebrand.

Sika deer (*Cervus nippon*). Photo courtesy of Dong Lei.

Eld's deer (*Rucervus eldii*). Photo by William McShea.

Female Père David's deer (*Elaphurus davidianus*). Photo courtesy of Li Sheng.

Philippine deer (*Rusa marianna*).
Photo courtesy of Freddy Pattiselanno.

Male (stag) fallow deer spar to establish a dominance hierarchy for breeding. Photo courtesy of Jackie Pringle.

whether this is a signal to the predator that the prey has detected its presence, a means of maintaining group cohesion, or a signal to other group members to flee; the flair signal may serve all of these purposes.

Most of the known social behaviors are for larger deer species. Smaller deer species tend to remain secretive, and researchers know little about their courtship or dominance behaviors, except what they observe in captive settings. We are aware, though, that the simpler structure of their antlers indicates fewer ritualistic interactions between males, and that none of these species form large social groups.

Do deer fight?

Yes, deer fight. Aggressive interactions are most common between adult males during the mating season, as well as among all adults, male and female, competing for access to resources. Both sexes will flail their hooves and butt heads during encounters over food sources in a limited area, such as bait piles, or when confined to a small space. Females can be aggressive with their offspring—either when their young insist on attempting to nurse or refuse to move from a bedding site (sleeping area)—and among adults at feeding sites where food is scarce, although this aggression is usually limited to a single kick or head butt.

During the mating season, fighting between males begins with a ritualized set of approach and engagement postures that then leads to the males locking antlers and sparring. The primary function of antlers—a defining characteristic of deer—is for ritualistic fighting between males to establish dominance hierarchies for mating and to convey their "fitness" to females. The branched structure allows males to engage their antlers and struggle for dominance with less serious injuries than would result from straight or spike antlers.

Sparring is often intense, however, and males can cause significant injuries to their opponents or suffer serious wounds themselves that can prove fatal. Their branched and complex antlers allow these animals to lock into each other, and sparring involves much pushing and twisting. Unmatched antlers, however, due to size differences or broken tines, cause the locked antlers to shift and potentially pierce combatants. For North American elk, it is estimated that 5 percent of the males die each year from wounds suffered during sparring.

The antlers of smaller species are not complex and may only form a spike, as in tufted deer or pudu. Because the antlers of competing males in these species do not fit together, sparring is rare, and the fights are more about thrusts, bites, and flailing hooves. In smaller species without elaborate antlers, such as Chinese water deer, the males have enlarged upper canines, and males and females will slash each other with their canines during fighting. These smaller species are more likely to compete over access to resources, however, than to mates.

Deer employ different behaviors to prevent or minimize fighting. As a social species, once dominance hierarchies have been established, subordinate animals (both male and female) will move aside in the presence of dominant individuals. The bluffing charges of males toward smaller bucks, the bugling and bellows of males, and the head-up, antler-back display of many New and Old World deer are displays meant to convey dominance and avoid actual fighting.

Most deer are not inherently aggressive. Deer can be kept in captivity, and many species have either been domesticated (reindeer) or farmed—for example, sika deer, fallow deer, and white-lipped deer (*Przewalskium albirostris*)—for centuries (see chapter 8). Smaller species that do not form herds have not been domesticated to any degree. For non-domesticated species that are raised in captivity, such as some white-tailed deer, a docile animal (particularly males) can turn aggressive during the mating season. Beyond castrating the animal, there is no way to prevent this onset of aggression.

How smart are deer?

The quick answer is that they are as smart as they have to be. For any large herbivore there are important processes that are critically important for their survival, such as avoiding predators, identifying edible food, recognizing social signals, and knowing their environment. Deer are good at

Red deer bugling during the breeding season (rut). Photo by Bill Ebbesen.

doing these things. Their ability to recognize threats and move away from potential predators, including humans, can appear intelligent, but these traits are probably the result of following simple and effective rules that are largely instinctual. Hunters frequently report seeing deer assessing the situation and appearing to make rational decisions. However, for every generalization about how deer respond to the sights, sounds, or scents of potentially dangerous incidents, there are deer that have an opposite behavior. This behavioral plasticity is an advantage for a prey species that is under strong selection pressure to outwit its predators.

Most deer have established home ranges, which means that they have a sense of place. White-tailed deer relocated up to 6 miles (10 km) from their home ranges are likely to return. Mule deer, red deer, white-tailed deer, and caribou may make extensive seasonal migrations, returning to the same location year after year. Deer change their approach to a feeding site or alter their daily movement in response to disturbances or harassment at a specific location. For example, deer become more active at night if there is increased hunting pressure during the day.

Mother-young interactions for white-tailed deer and fallow deer include recognition of specific individuals through vocalizations. Complex social behaviors during courtship and rut (mating), such as those for white-tailed deer and red deer, indicate a basic sense of planning, forethought, and anticipation.

The saying "like a deer in the headlights" suggests that deer often appear to act on instinct and do not realize the threat posed by rapidly moving or changing events. The frequency of collisions between deer and vehicles for North American and European species is one such case, where it appears that deer cannot learn to avoid roads with automobile traffic. The pattern of deer stopping in midroad, changing direction to head back across a road, and dashing across a busy road at the last second are traits familiar to many automobile drivers. Although these behaviors are not useful or adaptive around rapidly moving vehicles, they do make sense within the evolutionary context of a social species that must make rapid and erratic movements to evade the many carnivorous animals that prey on them.

Do deer play?

All mammals probably exhibit some degree of play, because it is a mechanism for practicing social interactions, testing muscle development, and coordinating responses to unexpected events. For deer, play is most often seen between young animals, or between mothers and their offspring. The young are usually the active play partners, jumping and running around the mother. Mothers are more tolerant participants than active play partners, but they will push their young with their heads. When twin fawns start following their mother, they can engage in jumping and nipping behaviors with each other. Encounters with novel animals, like frogs or turtles, will trigger jumping and pawing in fawns.

Yearling males will also engage in what should be considered play behavior with mature males. Spike bucks will challenge larger males to sparring bouts that the older male is in no danger of losing. These older males, like mothers, restrain themselves and appear to engage the youngster in mock battle.

Do deer talk?

Deer use vocalizations as an important means of communication. Deer vocalize around predators, during the mating season, and between mothers and their young. Most deer give a distinctive snort in response to perceived danger, and some have a particular bark in response to predators. Muntjac, also known as barking deer, give a loud, doglike bark when alarmed. Roe deer and Eld's deer also bark when in danger. Male sika deer make a noise similar to a high-pitched scream, while moose may roar when defending themselves or their young against predators. Caribou make a clicking sound as they walk, which is produced by a tendon rubbing across a bone spur in their feet. This clicking may serve to keep a herd together in times of low visibility. These sounds are not complex communication signals, but they do involve a standardized signal within the species that conveys a limited amount of information.

During the mating season, most verbal communication is between males, and these signals have more complexity. All deer species living in herds during the mating season have some sort of bark or loud call between competing males. Among red deer, North American elk, barasingha (*Rucervus duvaucelii*), fallow deer, and white-lipped deer, dominant males have distinctive bugles intended to indicate their strength. It is possible that these signals also convey the identity of the individual animal, but this has not yet been studied in detail.

Communication for all deer species includes some vocalizations between a mother and her young during the mother's subsequent return. Does make a bleating sound when approaching bedding sites, and fawns often reply with their own bleats. A fawn's distress or feeding call does not always solicit a response by its mother, but a mother's call to locate her fawn is more likely to produce a reply. A fawn can recognize its mother by her approach call.

Most deer also communicate non-verbally through chemical signals. Deer have an extensive array of glands on their limbs and around their head that excrete pheromones, which then are either dispersed into the air or deposited onto objects around them. All deer have an intraorbital gland used to anoint (leave scent on) branches and vegetation at head level. The deer places the tip of a branch within the gland—a tricky feat for an animal with no hands and an opening located very near the eye. Interdigital glands between the hooves deposit scent as the animal follows trails within its habitat. Metatarsal and tarsal glands on the hind legs also produce waxy, scent-laden secretions, particularly during the mating season. For marsh deer (*Blastocerus dichotomus*), glands on the back of their legs produce odors that are evident to other deer over 0.6 miles (1 km) away. Red deer, sambar, and barasingha will spray urine on their neck and head, and then rub these body parts on trees or along the ground. For most deer species, scientists do not know the specific function of each signal. They do know, however, that both males and females deposit scents extensively during the mating season. Small, territorial species like tufted deer, brocket deer, and muntjac deposit scent throughout the year by using scent posts. Latrines may also serve as scent posts. Captive male brown brocket deer (*Mazama gouazoubira*) maintain latrines within their territories that they countermark when feces from strange males are introduced.

The scents involved with mother-young bonding allow mothers to recognize their fawns. Researchers know little about the origin of scents being used as cues by mothers, but these females appear able to recognize their fawns both by smell and sound.

How do deer avoid predators?

Deer have few viable options when confronted by predators, who generally rely either on stealth (stalking) to approach and pounce on deer before they flee, or on endurance to outrun a deer (coursing). The primary defenses deer have against predators are their keen senses of smell, sight, and hearing. Visually, deer are quick to pick up movements, but they are limited in their ability to detect shape or color. White-tailed deer can see colors, especially in the lower (blue-green) wavelength, but their eyes are more suited for visual sensitivity and acuity (sharpness) under low-light conditions. A deer can detect objects over a wide field of view, about 310° around its body, because its eyes are set laterally (one on each side) on its head. White-tailed deer hear a range of frequencies that are audible to humans, in addition to sounds in the ultrasonic range, indicating that they can easily hear both predator and human movements.

Deer can be characterized as either "runners" or "slinkers" (hiders) in their behavioral response to predators. Deer whose primary predators are coursing species usually have some alarm call or activity that indicates when a predator has been detected. For most deer, their initial startle response is running. For white-tailed deer, this response includes flashing (flagging) their tail. This flagging involves raising a relatively long tail and erecting the hairs along the edges of a large white rump patch to increase the size and visibility of the patch. They also extend the hairs across their tarsal glands on the inside of the hind legs. In this context, the flagging behavior is thought to serve as a warning to other deer that danger is near. It also warns the predator that the deer are aware of its presence and that the opportunity to silently stalk them is past. White-tailed deer can also alert the herd by snorting and striking a hoof on the ground when something has aroused their suspicion. Other deer have similar responses to sighting a predator. For example, mule deer and pampas deer have a high-stepping alarm walk when fleeing a predator (known as stotting), similar to the stiff, four-legged bounce of fleeing antelope.

Following a startle response, some deer quickly transfer to a crouching run or retreat into the vegetation once they have lost sight of a predator. This behavior is more common in small species such as pudu, muntjac, tufted deer, and hog deer (*Axis porcinus*). Other small species, like brocket deer and huemul (genus *Hippocamelus*), respond to perceived danger by

Small species, such as this female hog deer, often retreat to the cover of vegetation. Photo courtesy of M. Monirul H. Khan.

Large species, such as these sambar stags, may stand and fight potential predators. Photo courtesy of Jackie Pringle.

freezing and remaining motionless. Some forest species, such as white-tailed deer, will run, but then they occasionally circle back behind the pursuing wolf pack or cougar. In this way, they do not leave their familiar home range. Hog deer and roe deer are amazingly good at dropping out of sight once they are flushed by a predator. In contrast, deer in more open habitats, such as caribou and chital, will run and keep running once they detect a predator. Eld's deer and pampas deer (*Ozotocerus bezoarticus*) are open-forest species that exhibit an initial startle run and then stop to look back; if the predator is still pursuing them, they will enter into a full run. Stopping to observe the predator might serve deer well against stalking enemies such as tigers, but not against human hunters using modern weapons.

Other deer responses to stalking predators may include erratic movements during their initial flight. Forest deer have trail systems that allow for rapid movement through dense vegetation. Moose are large enough to move over downed logs that predators may find difficult to leap.

Several species, such as marsh deer, sika deer, Chinese water deer, and moose, will readily take to water if pursued by predators. Standing and fighting against predators, however, is not a common response. Only the largest deer—moose, red deer, and sambar—show any propensity to fight. Usually it is a female defending her young who will stand and fight against a predator.

Most deer rely on concealment of their fawns until the latter are old enough to flee rapidly. Small deer—such as muntjac, brocket deer (genus *Mazama*), tufted deer, southern pudu, and hog deer—live in forested sites where the concealment of adults and their young is the first response to potential predators, regardless of the attack style of their enemies.

Chapter 5

Deer Ecology

Do deer sleep in the same place each night?

Forest deer do not sleep in a single location, but most have a limited number of sleeping areas, called bedding sites, within their home range. These sites most likely are determined by their degree of shade, the wind direction and speed, the slope and aspect of the ground, and their proximity to food sources. With several bedding sites in a home range, deer may shift which ones they use whenever their feeding sites shift. Female deer are particularly prone to using specific bedding sites during the fawning season. Deer species that inhabit open plains, such as Eld's deer (*Rucervus eldii*) or pampas deer (*Ozotoceros bezoarticus*), or reside in large social groups, such as chital (*Axis axis*) or migratory caribou (*Rangifer tarandus*), do not have regular bedding sites.

Most bedding sites are occupied by individual deer, but during the winter white-tailed deer (*Odocoileus virginianus*) in northern temperate forests may congregate in deer yards, where multiple deer will bed together. These mass bedding sites arise during times of deep snow cover and are found in areas with thermal cover that are protected from the wind, such as in stands of conifer trees. Deer either feed within the yard or regularly use trails leading from these bedding sites to other feeding locations, thus reducing the energy expended by individual animals to open up trails during heavy snows.

For non-migratory species, females will not bed with a newborn fawn, but rather will return to the fawn a few times a day to feed, groom, and possibly move it to a new location. If they have twins, deer will place each fawn in a different bedding site.

This stag chital, also called an axis deer, is alone, but the species often is found in large groups. Photo courtesy of M. Monirul H. Khan.

Regardless of the potential number of bedding sites, deer do not spend a lot of time sleeping. Because they are prey for many carnivores, deer are usually alert, even when bedded down. Species that have been studied spend about 4.5 hours per day sleeping and average only 30 minutes in a deep (REM) sleep state. Some researchers say drowsiness—being neither completely awake nor fully asleep—is a better description of how deer sleep. Deer do not limit their sleep to nighttime, but intersperse bedding with feeding and ruminating (re-chewing their food) throughout a 24-hour cycle. They lie down during sleep, as they lose muscle control and cannot lock their legs to remain standing, as elephants can. Thus bedding down usually serves a dual function, as a place to sleep and to re-chew their food.

Do deer migrate?

Although most deer species are sedentary throughout their adult life—meaning they remain within their home range—some migrate seasonally to different elevations, and a few species migrate thousands of miles each year. Moose (*Alces alces*) may migrate 190 miles (300 km) between their winter and summer ranges. The most nomadic species is the caribou of the subarctic plains and boreal (northern) forests. Some populations of woodland caribou are relatively sedentary, but they will continuously shift their home range throughout the year. For more northern populations of caribou, some radiocollared individuals in Alaska moved over 3,100 miles (5,000 km) in one year. Their movements are not necessarily to head north in the summer and south in the winter, but more often are determined by the location of snow-free areas in the winter and calving sites in the summer. Many of the physical and behavioral attributes of caribou are closely tied to the constant movement of individuals, in large herds, over vast open spaces with limited forage availability, while enduring harsh climates.

For elk (*Cervus elaphus*) and mule deer (*Odocoileus hemionus*) in the western United States, their seasonal movements are more likely to be along elevational gradients, with them living in high-elevation forests and fields during the summer and moving down to river valleys and lower pastures during the winter.

Which geographic regions have the most species of deer?

Deer occur naturally on all continents except Antarctica and Australia (see chapter 1). North and South America, as well as Eurasia, have a diverse collection of deer species. In North America, the moist forests and open plains of western Canada hold the largest assortment of deer—with white-tailed deer, mule deer, caribou, elk, and moose occupying the same

Large herds of caribou migrate in the Arctic. Photo courtesy of Anne Gunn.

region. In Eurasia the largest combinations of deer occur in South and Southeast Asia, where the ranges of chital, barasingha (*Rucervus duvaucelii*), hog deer (*Axis porcinus*), Eld's deer, sambar (*Rusa unicolor*), and several species of muntjac (genus *Muntiacus*) overlap. Large deer species are more often found in the boreal and temperate forests of both North America and Eurasia. The small species tend to live in the tropical forests of both South America and Asia.

The fewest number of deer species is found on the continents of Africa and Australia. In Africa, antelope have been the functional replacement for deer species, and they occupy all the ecological niches that deer use in Eurasia and the New World. There is one species, the water chevrotain or mouse deer (*Hyemoschus aquaticus*), which is found in equatorial Africa and looks similar to a deer, but taxonomists (scientists who classify organisms) place it in a different family (Tragulidae). Mouse deer do not possess antlers, but they have upper canine teeth that form tusks similar to those found in tufted deer (*Elaphodus cephalophus*) and Chinese water deer (*Hydropotes inermis*). Australia does not have any native deer species; the marsupial kangaroos and their relatives fill the niche of large herbivores in these ecosystems. However, earlier introductions have resulted in at least six species of deer presently found in captive and wild populations in Australia. In New Zealand, deer species are also extensively introduced on game farms and for sport hunting (see chapter 8).

How do deer survive in the desert or during droughts?

Some species, such as mule deer, can survive in a semi-desert habitat. Generally, however, deer are not as abundant in deserts as are other arid-region hoofed mammals (ungulates), such as camels, and bovids such as antelope. We know of no deer species that can survive for long periods without access to water. This is due in part to their relatively inefficient kidneys, which do not concentrate urine; deer lose more water in their urine than species adapted to very dry desert environments. Deer use behavioral modifications, such as shifting their feeding to nighttime, to increase the water uptake in their diet and reduce water loss through evaporation and transpiration (through the pores of the skin). Deer can reabsorb water through their large intestine and their colon, which is coiled. The removal of water as digestive waste materials move through this coiled colon results in deer depositing small, pelletlike feces. The tremendous expenditure of energy necessary for antler production in males and lactation (milk production) in females limits deer to habitats with extensive vegetation for at least part of the year. Deer are more common in savannas, where a mixture of woody species and grasses increases their access to food and water.

Muntjac are also called barking deer. The 11 species of muntjac are found in China and Southeast Asia. Photo courtesy of M. Monirul H. Khan.

In tropical regions, most deer survive in a monsoon weather pattern, which means heavy rains during part of the year and extended drought during the rest. The extent of the drought depends on the region, but in South and Southeast Asia, the usual dry period is six to eight months. During this time, little or no rainfall occurs. Deer enter the dry season drinking surface water within streams and ponds, and, as these sources evaporate, they restrict their movements to places with remaining surface water. Several species in tropical Asia, such as hog deer, Chinese water deer, and barasingha, confine themselves to riverine areas (on or near flood plains).

Two large deer species, Père David's deer (*Elaphurus davidianus*) and Schomburgk's deer (*Rucervus schomburgki*), were riverine species, but they were extirpated (completely eliminated) from the wild. One subspecies of Eld's deer, *Rucervus eldii eldii*, is confined to a single park in northeastern India, where a floating mat of vegetation, called fundi, sustains the entire subspecies. In South America, the habitat of marsh deer (*Blastocerus dichotomus*) is limited to open waters or marshes, and pampas deer occupy savannas with some surface water. For species that live outside of riverine areas, their home ranges still become restricted around open water sources, and they make extensive use of tree fruits during the dry season. Deer in seasonally dry habitats also feed more extensively in the mornings, to obtain moisture from the dew that forms on vegetation overnight.

How do deer survive the winter?

Generally, deer accumulate body fat during the summer and fall, and lose weight during the winter. For northern species, winter can challenge the physiological limits of deer: survival depends on the physical condition of the individual, the quantity and quality of available habitat, and the length and severity of the winter. Most deer are extremely flexible in their food preferences, so they can feed on grasses and forbs (any non-woody, broad-leafed plant that is not a grass) during the growing season, then switch primarily to the buds of woody species during the winter.

For temperate-climate deer species, the influx of hard nuts (called mast) in the autumn is important in building up sufficient fat stores. Deer in temperate, boreal, and Arctic regions rely primarily on these fat stores for survival over the winter. Deer must continue to eat during winter, regardless of their fat stores, because of the requirements of their digestive system. The rumen (the first stomach compartment) of a deer contains bacteria and protozoans, microorganisms that digest plant fiber into its simpler parts of sugars, proteins, and amino acids. These microfauna must be nourished on a regular basis. Poor quality forage, including strips of bark and dead vegetation that has little nutritional value, may still serve a valuable function in maintaining the necessary materials for this microbial activity.

Most winter deer deaths (mortality) occur when their fat reserves are used up but new plant growth has not yet started. Overwinter survival is primarily a function of body size. Larger deer, with their favorable surface-to-volume ratio, have greater fat stores and expend less energy per ounce of body weight then do smaller deer. For this reason, growing fawns that do not reach sufficient body size during the summer and fall are not likely to survive the winter. White-tailed deer fawns enhance their ability to survive by undergoing lipogenesis in the fall, where fat stores are accumulated no matter what the level of food availability. While these stored fats extend their chances of survival, over the winter season fawns still experience the highest death rate among age classes of deer. Winter mortality also includes adults who did not store sufficient fat for their body size and that winter's severity. This may occur in populations occupying poor quality habitats, or whose numbers exceed the carrying capacity of their habitat—that is, the ability of the land to contain enough resources to support them. Within these populations, males that were exhausted or wounded in the autumn rut (mating season), injured and sick animals that were not able to eat enough during the fall, and older individuals whose teeth are excessively worn are all susceptible to starvation. A third source of winter mortality for temperate and boreal species of deer is increased predation. Heavy snows limit the movement of deer and make them more vulnerable to predators such as gray wolves (*Canis lupus*).

In the winter, many temperate-zone species, such as white-tailed deer, congregate in communal sites called deer yards (see discussion on "sleeping" on page 55). These yards have reduced snow depths and radiating trails; are usually associated with conifer cover, which provides protection from snowfall and from predators; and allow deer to move more easily. Caribou survive harsh winters by moving long distances to find snow-free areas where they can forage on moss and lichens. Most species, including caribou, use their hooves to scrape away snow. It is a common misconception that the enlarged, asymmetrical brow tine over the face of caribou (called the "digger") is used to scrape away snow. Males drop their antlers before winter, so the enlarged brow tine most likely has a role in courtship behavior. Mule deer, elk, and red deer (*Cervus elaphus*) will move to lower elevations or snow-free slopes, where they have more access to winter forage. Moose feed primarily on the winter buds of small trees, and they live in marshy areas that experience relatively longer growing seasons.

Do deer have enemies?

The primary predators of deer hunt them either by stalking or coursing (running). Stalking predators, like mountain lions (*Puma concolor*) and leopards (*Panthera pardus*), rely on stealth to approach bedding and feeding sites and pounce on deer before they flee. These enemies are capable of short, rapid chases, but they cannot sustain a chase beyond 30 to 60 seconds. They kill usually by leaping onto the back of the deer, with a crushing bite applied to the neck. The ultimate stealth predators are constrictor snakes (i.e., boas, pythons, and anacondas) that are capable of eating a small deer if it can be captured at a bedding site.

Large, slow predators, such as bears (family Ursidae), or smaller predators, such as foxes (genus *Vulpes*), coyotes (*Canis latrans*), and jackals (genus *Canis*), prey primarily on fawns found at bedding sites. These predators are rarely able to capture an adult deer, unless its movement is severely hampered.

Coursing predators rely on endurance to outrun a deer. These predators are larger canids, such as gray wolves, dholes (*Cuon alpines*), and domestic dogs. They rely on persistence to run down a deer, and hunt in packs; different pack members will trade off pursuing a single deer. In the final moments of the chase, the predator will either grab the muzzle of the deer, disembowel the animal by biting at its underbelly, or disable it by cutting a hamstring muscle in one of its back legs. There is evidence that these predators can select individuals from a herd that appear weak or injured. They may also choose young animals that are not as fleet or as dangerous as adults.

Deer have few viable options when confronted by predators (see chapter 4). Most deer rely primarily on rapid running. Only the largest deer, such as moose, red deer or sambar, may try to defend their young or themselves from attack. Most deer in North America and Europe are killed by humans, either through hunting (harvesting) or collisions with vehicles (see chapter 9). Worldwide, less than 10 species of deer are legally hunted. It is difficult to estimate the amount of illegal poaching of rare species, but the smaller species of deer in Asia and South America are a significant part of the bushmeat found in local commercial markets. Larger species are also legally harvested in significant numbers. In 2008, the most recent year with complete tallies, over 6.8 million white-tailed deer were harvested in the United States, with 45 percent of these animals being antlered males. This annual harvest represents about one quarter of the total deer population. It is estimated that more than 1.5 million deer-vehicle collisions occur in the United States each year. These collisions are usually fatal for the deer, and they result in $1 billion in damages to automobiles and 30,000 human injuries. Moreover, this is an issue that reaches beyond problems with white-tailed deer in the United States. In Sweden, deer-automobile accidents involve at least 10,000 moose and 50,000 roe deer each year. These deer deaths far exceed those due to predation or disease among populations of white-tailed or roe deer.

Do deer get sick?

Disease in deer falls into three broad categories: viral, bacterial, and parasitic (i.e., caused by parasites). All types of deer diseases are readily transferrable, and they are more common where the population densities are high. Most of what we know about disease in deer is from studies on North American and European species. Because deer are related to livestock, such as cattle and goats, some diseases are transferred between them. Because living conditions for livestock are usually at much higher densities than those for wildlife, unvaccinated livestock are more likely to transfer disease to wildlife than vice versa. Researchers suspect that the numbers of some endangered deer in South America, such as southern huemul (*Hippocamelus bisulcus*) and taruca, or northern huemul (*Hippocamelus antisensis*), are declining because of diseases transmitted by livestock.

Bacterial diseases among deer are common; the ones most studied in North America are bovine tuberculosis (bovine TB, or *Mycobacterium bovis*) and brucellosis (*Brucella abortus*). Both diseases can be passed from one deer

A collision between a white-tailed deer and a vehicle in Virginia.
Photo background modified from www.outdooroddities.com.

species to another, and between deer and domestic livestock. Bovine TB probably was originally transmitted from cattle into elk and white-tailed deer populations, and it is now present in isolated deer populations in the upper Midwest and the western United States. Brucellosis is currently a serious problem only in elk herds in the Greater Yellowstone Ecosystem. Some bacterial diseases, like bovine TB, are transmitted as an aerosol (droplets suspended in the air) and others, like brucellosis, need physical contact with a contaminated article.

Sickness in deer frequently results from parasites. Most parasites do not cause death unless the population densities of the deer are high or individual animals are otherwise stressed. Large stomach- and lungworms occur frequently in all types of deer that have been studied. Common parasites in North America are nematodes (roundworms), such as meningeal worms and muscle worms. Some nematodes require an intermediate host—usually gastropods, such as snails—that pick up the worm larvae while consuming deer feces.

Liver flukes are a type of flatworm parasite found in deer. They depend on an aquatic gastropod during their larval stage, and thus are confined to species that forage in aquatic areas, such as moose. Arterial worms, transmitted by horse flies, are common in the blood systems of mule deer, white-tailed deer, and elk. Typically, there are few outward indications of parasite loads (the amount of parasites in an individual's system). On the other hand, at higher levels of parasite loads, deer can appear lethargic (sluggish, or lacking energy), with low body weights and loose stools.

In the southeastern United States, management agencies may use the number of nematodes in samples from the true stomach (the abomasum) of white-tailed deer as an indicator of herd health. This is called the Abomasal Parasite Count (APC). The higher the APC, the greater the probability that other diseases and malnutrition will cause deer deaths. Deer differ in their response to specific parasites. Meningeal worm is chronic (frequently recurring) and widespread in white-tailed deer, for example, but doesn't cause death; yet it is fatal to moose and other cervids. This difference in susceptibility may restrict the overlap in the geographical distribution of white-tailed deer and moose.

Common viral diseases are bluetongue and the various hemorrhagic diseases (those that cause significant bleeding), which include epizootic hemorrhagic disease. Both are transmitted to deer from biting midges of the genus *Culicoides*. These diseases usually appear in late summer and early fall. Infected deer may exhibit lethargic behavior, low body weights, and excessive thirst. Outbreaks of these diseases can cause a significant number of deaths and result in noticeable reductions in population density.

A newly recognized and serious disease for deer is chronic wasting disease (CWD), which is a transmissible spongiform encephalopathy, or prion disease (see chapter 9). CWD is part of a family of diseases, including sheep scrapies, mad cow disease, and Creutzfeldt-Jakob disease in humans. Prions are persistent, abnormal (aberrant) proteins that can be transmitted among deer through contact with the body fluids of infected animals, or with soils and other materials containing these fluids. In North America, CWD has been found in elk, white-tailed deer, and mule deer. It was originally identified in the western United States, but it has now spread to midwestern and eastern states, probably through the transport of captive deer. This disease affects the brain and central nervous system: infected deer exhibit excessive thirst, hypersalivation (the production of too much saliva), disoriented behavior, weight loss, lethargic movements, and eventually death.

The prion diseases do seem to be specific to individual species (i.e., species specific), so that there is no evidence of CWD spreading from deer to humans. Nonetheless, the Centers for Disease Control and Prevention recommends that people avoid handling blood and marrow products from infected deer. At present there are no means of controlling the spread of CWD beyond drastically lowering the density of deer populations through culling (removing selected animals from the herd, usually by professional sharpshooters)—which is what Wisconsin and Illinois have attempted.

Generally, a deer suffering from disease may have a rough coat, closed eyes, yellow discharge around its eyes and nostrils, and loose stools. If a disease causes the deer to have a lack of appetite or an inability to move about and forage, the animal loses weight, exhibiting a visible rib cage and shrunken haunches. Most deer store body fat in their haunches; loss of this fat will result in a hollow appearance. A lack of proper mineral and vitamin

An elk in the last stages of chronic wasting disease. Photo by Terry Kreeger, Wyoming Game and Fish Department.

uptake can cause several skeletal problems in deer, including antler breakage, deformities in the developing bone structure of young deer, and reduced bone density (called osteomalacia).

Deer are hosts for a number of species of ticks, which concentrate primarily around the deer's head and ears. Deer are an intermediate-stage host for many tick species; 18 species of ticks are reported for deer in North America alone. Heavy tick loads are unsightly, but generally they do not represent serious harm to the health of adult deer. One exception is in northern populations of moose, where infestations of up to 30,000 ticks on one individual can cause hair loss, an inability to maintain body heat, debilitation, and eventual death. Large numbers of ticks on fawns can also result in death.

Sick deer should not be handled or rescued, except by professionals. This is not due to the danger of transmitting disease from deer to humans, but rather is necessary for the safety of the deer and the persons trying to handle a wild animal. Even sick deer can use their hooves to injure people.

It is common for the public to assume that rescued fawns are sick. Fawns do not have much body fat and are generally spindly and thin. Within the first few days of their birth, fawns will freeze if they are discovered and will appear docile and weak. It is difficult to judge the health status of a fawn, and finding one without its mother nearby is not evidence of a problem (see chapter 8).

Are deer good for the environment?

Deer play an essential role in the environment, both because of the foods they select and because of the food they provide for other animals. Deer convert plant material into the protein and carbohydrates used for their own growth and reproduction. Thus, they continue the conversion of sunlight into usable energy by the various living components of the community within which they reside. Deer can change the relative abundance of plants through their selection of vegetation that is high in energy and low in plant defense mechanisms (such as spines, thick skins, or toxins). This selection process is part of the natural cycling of energy and nutrients. Lovers of spring wildflowers may strongly disapprove of deer who munch through their favorite flowers, but the fact remains that wolves and other carnivores (meat-eaters) cannot digest flowers; they need plant energy converted into an edible form to survive—and deer serve that purpose. In other words, a functioning ecosystem needs all its parts: plants, herbivores, predators, and decomposers. However, overabundant deer populations can cause problems for natural-resource managers, as is the case with white-tailed deer in some areas.

In many ecosystems, deer are a critical resource for large predators. Studies within Asian forests have found that deer comprise up to 80 percent of the diet of top predators like tigers (*Panthera tigris*) and leopards. Wildlife managers in Asia use the density of large herbivores (of which deer are the primary species) as a measure of how many predators a park can support. In North America, wolves rely primarily on elk, mule deer, white-tailed deer, caribou, and moose for their nourishment. Mule deer are the main food source for mountain lions in wilderness areas in the western United States. Even the diversity of deer is important to many ecosystems. Young tigers in Asian forests may feed mostly on smaller deer species such as muntjac, chital, and hog deer; then progress to medium-sized prey like wild boar (genus *Sus*) and Eld's deer; before being able to capture and kill sambar or large bovids, such as gaur (*Bos gaurus*) or banteng (*Bos javanicus*).

Deer feces and urine contain nutrients and minerals that are quickly used by the microbial community and reabsorbed by plants. Deer that feed in fields but bed in forest sites can move significant amounts of nutrients from fields to forest systems by defecating in areas apart from their forage sites. Scavenger birds and other animals feed on deer carcasses, which also provide necessary food for carrion insects. Animals, including other deer, consume the bones of dead deer—they need calcium for their own bone and antler growth. In short, deer are certainly an integral part of the environments in which they live.

Chapter 6

Reproduction and Development

How do deer reproduce?

Reproduction in deer is a complex process where males and females must be synchronized in their hormonal and physical development. Sperm must be deposited within the reproductive tract of the female within hours of ovulation (production of an egg from a mature follicle—a group of cells—in the ovary). The signal triggering this mutual reproductive readiness between individuals is tied to hormonal changes, which, in turn, are keyed to increasing or decreasing day length—known as a photoperiod (the number of hours between sunrise and sunset)—and is tracked by the production of the hormone prolactin in the pituitary gland.

Reproductive physiology (the functioning of the reproductive system) is controlled by hormones produced by the pituitary gland and the gonads. The production of a follicle-stimulating hormone from the pituitary gland promotes the development of sperm in the testes of males. In females, a complex sequence of hormone production—follicle-stimulating hormone, luteinizing hormone, and progesterone—results in ovulation of an egg. Along with egg and sperm development, both genders have secondary sex characteristics that must develop to ensure mating. For male deer, testosterone secretion triggers antler growth, increased neck musculature, testicular growth, and pheromone secretion (the release of chemicals that trigger a social, often sexual, response) from glands. Secondary sex characteristics are more pronounced in large deer species that are found in open habitats, such as red deer (*Cervus elaphus*), white-lipped deer (*Przewalskium albirostris*), Père David's deer (*Elaphurus davidianus*), Eld's deer (*Rucervus eldii*), and barasingha (*Rucervus duvaucelii*).

For most animals, their body organs do not increase proportionately with their body size; larger animals have relatively smaller organs. This is not the case with deer antlers, however—larger species have correspondingly larger antlers. The largest antlers for their body weight are in caribou (*Rangifer tarandus*); the largest overall antler size occurs in mature moose (*Alces alces*). Scientists do not know why large deer have proportionately larger antlers, but it may be because these species tend to occupy more open habitats and engage more often in rutting behavior during the mating season.

Because males may copulate with multiple females, they will produce sperm over an extended period that coincides with ovulation in all or most females within the population. For some species, such as pampas deer (*Ozotoceros bezoarticus*) and muntjac (genus *Muntiacus*), some males in a population will have viable sperm (capable of impregnating females) throughout the year. For individual males, sperm production is limited to a specific period that is synchronized with the antler cycle, although these two physical aspects may be driven by different hormones. Sperm development and testosterone production respond to an increase in follicle-stimulating hormone; increased testosterone coincides with antler growth from a bony pedicle (an extension of the frontal bone) on the skull of males. The antler is composed of bone that grows under a network of blood vessels and highly nerve-filled tissue, called velvet (see chapter 2). The velvet is extremely sensitive and easily damaged, and male deer do not engage in any sparring or fighting when their antlers are in velvet. Rising testosterone levels, however, cause the velvet to slough off and expose the hardened bone, so males can then use their antlers for sparring. After the rut (mating season), and a decrease in testosterone, deer shed their antlers, which also coincides with a decline in sperm production. Male deer do not produce sperm again until there is a rise in their pituitary hormones, which triggers a renewed production of testosterone by the testes, and the antler cycle begins again. For most species of deer this is an annual cycle, with a single mating season.

The maturation of eggs within a female's ovaries is also triggered by an increasing production of pituitary hormones, including follicle-stimulating hormone and luteinizing hormone. Female breeding behavior is called estrus (commonly referred to as "heat"), and it is driven by estrogen produced by the maturing follicles in the ovary. Estrous behavior can include seeking out congregations of animals; excreting pheromones through vaginal secretions and urine, and a willingness to stand for mounting (sexual contact) by the male. The range of behaviors is specific to each species. Female white-tailed deer (*Odocoileus virginianus*), for example, dramatically increase their movements during estrus; female moose do not. Estrous be-

An Eld's deer with its velvet stripping (or in "tatters") during the spring.

Photo courtesy of Lisa Ware.

havior may last one to two days. If fertilization occurs, a female does not enter estrus again for that breeding season. If fertilization does not occur, a female will recycle and enter estrus at 18- to 30-day intervals, depending on the species. Ovulation without overt estrous behavior (silent heat) has been documented in caribou and other species of deer.

Males and females generally do not mate until the year following their birth (when they are yearlings—1.5 years of age), but there are exceptions. Small deer, such as Chinese water deer (*Hydropotes inermis*), can reach sexual maturity at 5 to 7 months of age. From 30 to 50 percent of female white-tailed deer living in good habitat in the southern and midwestern United States can become pregnant as fawns (at 7 months of age). Male white-tailed deer can produce "button" antlers during their first year, but in the wild these animals do not engage in mating behavior. Although physiologically capable of breeding, young males of many species are excluded from rutting by dominant older bucks. For example, male fallow deer (*Dama dama*) in the wild may not breed until they are 4 years of age.

Gestation length (the time from conception to birth) is generally related to body size, with smaller species, such as muntjac, having a gestation period of about 90 days, while gestation in Père David's deer is 290 days. A notable exception to the relationship between body size and the length of gestation occurs in European roe deer (*Capreolus capreolus*). The time between mating and birth in this small (66 lb, or 30 kg) species is 300 days. This is misleading, however, because roe deer exhibit delayed implantation—the fertilized egg floats free in the uterus for about 120 days before attaching itself to the uterine wall. Thus, actual gestation is only about 6 months. Roe deer are the only species of deer known to have delayed implantation. For some species, such as moose, fetal growth rate during gestation is linear. For other species, like white-tailed deer, the growth rate of the fetus is closer to exponential (a J-shaped curve), with most of its mass being accumulated during the last weeks of gestation. This allows a female to maintain her mobility for an extended period and delays the significant nutritional demands of a large fetus.

When do deer mate and give birth?

The timing of female reproduction is determined by environmental conditions that will be favorable for fawn development, such as the plant-growth season, which is determined largely by temperature in temperate zones and rainfall in the tropics. A relatively fixed gestation rate, based on body size, means that the optimal times for birth and lactation (milk production) result in a restricted period for mating and conception. For species in temperate regions, a birthing season in late spring coincides with a mating season in autumn. For example, most births for white-tailed deer in the mid-Atlantic United States occur from mid-May until mid-June. Given the necessary gestation period, most mating therefore occurs from early to late November. Extensions of the mating season are usually due to females that failed to conceive during their first estrus. A restricted mating season (in temperate regions) means that all males are on a similar annual cycle of antler development, coming into their hard antlers at approximately the same time. In the Northern Hemisphere, decreasing day length after June triggers changes in the gonads and the pituitary gland that then set the reproductive cycle in motion. The role of photoperiod is evident in white-tailed deer, which have a broad latitudinal distribution; the peak breeding season for populations in central North America is November, while April is the peak breeding season for populations in South America. South American species transferred to zoos in North America will shift their breeding season accordingly.

For tropical species, variability in the amount and timing of annual rainfall results in a mixture of strategies for breeding and birthing seasons. Some species have little evidence of fixed reproductive seasons, such as muntjac or brocket deer (genus *Mazama*), while others have birthing seasons that show one annual peak, as in marsh deer (*Blastocerus dichotomus*), or two annual peaks, as in pudu (genus *Pudu*). Eld's deer is an example of a tropical species with a constrained breeding season. Their reproductive cycle is set by photoperiod, even with the very subtle changes in day length found in tropical latitudes.

Do deer breed only one time per year?

Female deer only reproduce once a year. For females, a single breeding season is dictated by the time demands of gestation and lactation. For tropical species such as brocket deer, muntjac, marsh deer, and pudu, researchers have observed a postpartum estrus (breeding soon after birth of a fawn). Although breeding rapidly follows birth with these species, the long gestation period (approximately 200 days) establishes a cycle that approaches a single birth a year.

How many fawns do deer have?

Most mature female deer have a single fawn per year. Especially with the larger deer species, even one fawn or calf represents a large energy investment. There are, however, more fecund species. Among larger deer, a female white-tailed deer will produce two to four fawns annually when living in good habitat. Among smaller species, Chinese water deer are the most productive, with up to five fawns.

For an adult female, fawn production is relatively constant throughout her life. This means that, for most deer species, each female can produce from 10 to 15 offspring during her lifetime, depending on longevity. A red deer study in Scotland, for example, found that the average hind (female) has an 11-year reproductive period and produces approximately seven calves, with two surviving to adulthood. In contrast, a female white-tailed deer can produce close to 30 fawns during her lifetime.

For a male, the number of offspring sired during a single season, as well as throughout his lifetime, is more variable. For example, a one-year study of white-tailed deer in Michigan found that only 20 percent of males 3 years of age or less fathered offspring. Males more than 3 years of age fathered one to nine offspring. However, studies of different populations of white-tailed deer have produced varying results. An extreme example of male fecundity is a male (stag) red deer in Scotland that sired 94 offspring during his lifetime.

Are all deer fawns full siblings?

Based on genetic analyses of white-tailed deer and roe deer, about 20 percent of the fawns in multiple births are not full siblings. For these and other species, females mate with more than one male during their estrous period. For deer that congregate during the mating season, the traditional assumption was that the largest male was the only one fertilizing each female within the social group—and that the other observed matings were outside the peak fertility period. This assumption no longer appears to be true. Researchers, however, do not know how common multiple paternity is for smaller deer species that exhibit more pair bonding during the mating season, such as muntjac, pudu, and brocket deer.

Can the sex of a deer be determined visually?

Visual determination of gender in deer is often possible, simply because males grow antlers and females do not. The exceptions are caribou, where both males and females have antlers (although the antlers on males are larger and more elaborate), and Chinese water deer, where neither sex has antlers. For tufted deer (*Elaphodus cephalophus*), the antlers are so small in the males that they are rarely visible. Using antlers as a guide to gender will work for most deer species during the mating season. Because most species do not mate throughout the year, however, an absence of antlers does not always indicate a female. If a closer look is possible, you may see pedicles on males—the place where the antlers originate (although both genders have pedicles in muntjac)—or a visible penis and testicles. Individual species also have more specific visual cues differentiating males and females. Female moose, for example, have a white vulval patch on their genital organs, and male mule deer (*Odocoileus hemionus*) have a broader hoof print than females. In examining skeletons, adult male skulls can be identified by the pedicles, and the pelvic bones in some species (including whitetails and mule deer) have tuberosities (small projections) where the penis ligaments were attached.

Do deer care for their young?

All female deer provide care for their young. Lactation is energetically demanding for the doe—nursing results in a higher energy expenditure for females than gestation—but it is essential for the survival of fawns. In addition to nutrients, colostrum in the initial nursing milk contains antibodies that help with disease resistance. Mothers also groom the new fawns, and stimulate urination and defecation by licking the anal area of their young.

A male tufted deer with obscured antlers and an obvious upper canine tooth. Photo by Heush.

By licking around the mouth of her fawn, a doe supplies essential rumen microbes, necessary in deer digestive systems, through oral transfer to her young (see chapter 7), although these microbes can also be picked up from the environment.

Mothers also provide their young with some protection against enemies. In larger species such as moose, caribou, white-lipped deer, and red deer or elk, mothers can often defend their young against predators. For nomadic species, such as caribou, the calves must start following their mothers within hours of their birth. For smaller species, there is little that a mother can do against a larger predator beyond hiding her young. For these "hider" species, the mother leaves her newborn in one place and forages for food elsewhere during most of the day. Good examples of hider species are white-tailed deer, hog deer (*Axis porcinus*), and huemul (genus *Hippocamelus*).

The maternal behavior of white-tailed deer has been well studied. The young will be nursed only once or twice a day; the mother will return to near where the fawn was left and emit a call, to which the fawn replies. Fawns, as they develop, will begin to voice their own call to initiate nursing. Fawns also have an alarm call that will bring the doe to them. Mothers

A young red deer calf nursing. Photo by BS Thurner Hof.

occasionally move their fawns following a nursing session, but rarely more than 100 yards from the birthing site. When twins or triplets are produced, the mother will hide them in separate locations. This pattern will continue for their first six to eight weeks of life, until the fawns start following their mother as she forages. For all deer, once the young start following their mothers, they learn travel routes, feeding areas, and alert behavior from her and from other individuals in the herd or in the matriarchal group.

Researchers have observed fawns nursing from does that are not their mothers (called "allosuckling"); this occurs in species that form herds, such as caribou, roe deer, pampas deer, and red deer, as well as white-tailed deer. In no species, however, is allosuckling a common occurrence.

For all deer species that have been studied closely, males provide no direct care for the young. Male caribou and red deer will defend the herd or their harem against predators, but not a specific calf. For species in open habitats, males in mixed-gender herds do provide predator-alert signals or calls that benefit the entire herd, obviously including their own offspring. Researchers know little about the maternal and paternal behavior of the smaller species of forest deer, and it is possible that males in these pair-bonded species provide more direct care for their young.

How fast do deer grow?

Growth rates depend on a number of variables, including gender, population density, and habitat conditions. Nonetheless, the growth rates of deer are remarkably similar among species. Deer grow rapidly during gestation and lactation, but, once weaned, their growth slows. Birth weights

A fallow deer female (hind) grooms her young as they nurse. Photo courtesy of Jackie Pringle.

are approximately 5 to 10 percent of adult female weight. Prior to weaning, fawns and calves gain about 1 to 2 percent of their body weight daily when forage conditions are good for their mothers. At weaning, the young have gained 50 to 60 percent of their adult body weight; lower percentages occur in the largest species, such as moose and elk. Weight at weaning is important for temperate-region and boreal (northern) species, where the overwinter survival of the young is largely determined by body size. For most species, newborn males grow faster than females. In yearling mule deer, for example, the average female's body weight is 10 percent lower than that of a male. Similar differences in size and weight (sexual dimorphism) have been documented in sika deer, fallow deer, and most other species. Once an animal is weaned, its growth rate declines and becomes somewhat cyclical. Adults can reach their optimal, or maximum, weights by the end of their second or third year, but annual fluctuations occur because of reduced forage in the winter or during dry seasons, as well as energy expenditures due to the demands of reproduction and the maintenance of body temperature. Body weight usually declines in old deer, due to worn teeth and poor digestion of food.

How can you tell the age of a deer?

There is no exact visual measure of a deer's age. For larger species, antler size and configuration, along with body measurements, can be used to roughly estimate a deer's age as being young, adult, or old adult. Antler size, and (for some species) the number of tines (points), will increase as an individual grows older—although the number of tines does *not* correspond to an animal's age in years. This increase is mostly a function of greater body size, and, once an animal reaches the maximum size for the species, its antler size stays relatively constant. When the body condition of an old animal declines, due either to poor forage or an inability to chew food properly because of worn teeth, antler size decreases as well. For a well-studied species, such as white-tailed deer, it is possible to use online programs to estimate an animal's age using antler and body-size measurements. The probable age range derived from these programs can be rather broad, and it is influenced by habitat quality. No such online programs have been created for most other deer populations, however, and this would be difficult for the smaller species, such as pudu or tufted deer, whose antlers are a single spike or a forked tine.

To estimate the age of both male and female deer, most biologists use characteristics of their dentition (teeth). For white-tailed deer, mule deer, elk, moose, sika deer, red deer, and other species, the eruption pattern and wear of the molars and premolars (called cheek teeth) in the lower jaw give a good estimate of a deer's age through the animal's first 6 years. Beyond 6 years of age, the variation in the amount of tooth wear makes estimating age very unreliable. Moreover, intensely harvested deer populations rarely have individuals that live more than 6 years—and most management agencies simply group statistics for deer that are 4.5 years of age and older. There is also some variation in tooth wear among populations, based on the amount of grit in their diet, so that site-specific charts are necessary for precise age estimates.

It is also possible to estimate the age of deer through the annual layers of cementum (a specialized calcified substance) deposited on the roots of their teeth. These layers can be stained and counted under a microscope, in the same way that you would count growth rings on a tree. These lines

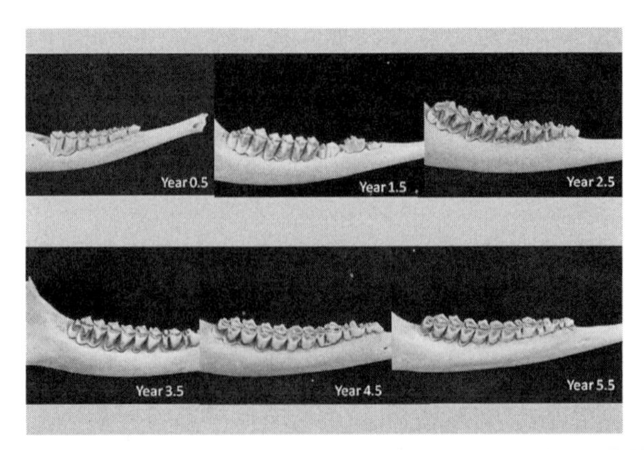

A comparison of increasing wear (from left to right and top to bottom) on the cheekteeth (premolars and molars) of an adult white-tailed deer.
Photo courtesy of Chad Stewart.

(called *cementum annuli*) are easier to count and are more reliable for temperate-region and boreal deer species than for tropical species, because the tropical species do not have distinct annual growth periods.

How long do deer live?

For most deer species, our only records of longevity come from captive populations. Life span is somewhat related to body size, with larger deer species living longer than smaller species. However, differences in longevity are not as great as might be expected; ages in smaller species, such as tufted deer (12 years), muntjac (up to 19 years), and pudu (up to 19 years) almost match the longevity of larger species, such as moose and elk (both recorded at 27 years). In the wild, however, most individuals live nowhere near this long—that is, there is a significant difference between potential (or physiological) longevity, and realized (or ecological) longevity. The mortality rate for wild populations is a U-shaped curve—with many fawns dying before reaching 1 year of age, followed by a relatively low rate of mortality during the prime of life, and then a high death rate in older deer. Mortality in fawns is usually due to predation or malnutrition during their first winter. In older deer, their molars wear down and food processing becomes less efficient, resulting in decreased body condition.

For hunted species, such as white-tailed deer, most males are harvested by 5 years of age, and very few females reach 8 years old. Within nonhunted populations, accidental death, predation, and disease kill most deer before age 10. For moose, another well-studied species, a 20-year study on Isle Royal in Michigan found that for some cohorts (i.e., animals born in the same year), only 10 percent survived to 5 years of age, while in other cohorts, 10 percent lived to age 10.

Chapter 7

Foods and Feeding

What do deer eat?

All deer are herbivores, and their diet consists almost entirely of plants. Their teeth and digestive systems are specifically adapted to cut and grind, digest, and then process nutrients from many different types of vegetation. Herbaceous plants provide a portion of the diet for all deer species, but the feeding strategies of different deer species are usually placed along a continuum, ranging from grazers (feeding primarily on grasses) to browsers (largely consuming the leaves and buds of woody plants). The position of a deer species in this progression depends on the availability of these different plants in their particular habitat and on each species' morphological adaptations (i.e., evolved physical characteristics). Open-habitat, large-bodied deer tend to be grazers, while forest-dwelling, small-bodied species tend to be browsers that select specific plant parts. There are several important exceptions to this pattern: large-bodied moose (*Alces alces*) browse on woody and aquatic vegetation; and two smaller species, hog deer (*Axis porcinus*) and chital (*Axis axis*), graze on grasses in floodplains.

Plants differ in their palatability (the acceptability of their taste) and digestibility, as well as the degree of grinding or chewing needed prior to digestion. Herbaceous plants with succulent (fleshy and moisture-filled) leaves and flowers are relatively easy to digest and provide the greatest amount of nutrition for their weight. Most deer species prefer to eat herbaceous plants; smaller deer will feed almost exclusively on succulent plants or buds within their home ranges. These latter types of deer—referred to as concentrate feeders—are very selective about both the species of the plant and the plant parts that they consume. Aquatic plants are one cate-

Several deer species form large groups when in the open, such as these red deer in Germany, grazing in a field.
Photo by Traroth.

gory of succulents, and several deer species—including moose, marsh deer (*Blastocerus dichotomus*), barasingha (*Rucervus duvaucelii*), and Chinese water deer (*Hydropotes inermis*)—specialize in feeding on aquatic plants growing in shallow waters or along the edges of ponds and streams.

Relative to other types of vegetation, grasses are low in nutritional content and contain a high percentage of non-digestible materials, such as silica. Silica, lignin, and cellulose in the cell walls of plants provide support and allow the plants to stand upright, but mammals can't digest these materials. Although many open-habitat species will consume grasses, no deer species feeds exclusively on them. Unlike the more sizeable species of bovids (such as antelope or bison)—which have large, heavy molars—a deer's teeth do not have the grinding force necessary to break apart the silica-laden cell walls of mature grass plants. The larger rumen (the first stomach compartment) commonly found in massive bovids can also digest grasses more effectively. An exception for deer occurs in springtime in temperate zones, when grass is in its early growth stage; or follows the onset of monsoon rains or fires in tropical forests and savannas, conditions which again encourage new grass growth. At these times, grasses offer deer an abundant, easily digestible food source.

Species of deer found in grasslands—pampas deer (*Ozotoceros bezoarticus*), hog deer, mule deer (*Odocoileus hemionus*), chital, fallow deer (*Dama dama*), and Eld's deer (*Rucervus eldii*)—can also take advantage of the food preferences of cattle and wild bovids (like bison and antelope), which crop mature grasses, allowing deer to concentrate on the succulent regrowth. In North America and Europe, most of the deer feeding in pastures or hayfields during the summer and fall are selecting succulent forbs (broad-

A moose feeding in a shallow lake in Alaska. Photo by Kate Banish, USFWS.

leafed plants) from among the blades of grass. In the spring, they are eating the new grasses. Because rice is a domesticated form of grass, Asian grass-eating deer, such as hog deer, Chinese water deer, Père David's deer (*Elaphurus davidianus*), and Eld's deer, will readily consume rice crops—inevitably bringing these deer into conflict with farmers, who regard them as serious pests.

Some deer species forage on the succulent portions of woody plants, such as buds and new leaves. The larger the deer species, the larger the size of the stem they can consume to extract its nutrients, allowing moose and sambar (*Cervus unicolor*) to make use of a wider range of woody plants than can the smaller muntjac (genus *Muntiacus*) or brocket deer (genus *Mazama*). In addition, taller deer species, like moose, have access to more terminal (upper) branches of an individual tree. As a tree grows, many of its succulent parts grow beyond the range of browsing deer, although deer can stand on their hind legs to reach forage on higher branches. Larger deer also can respond by breaking off the top of the tree. For all forest deer populations, when their numbers are more highly concentrated, browsing on the lower limbs of conifers (evergreens) or deciduous trees (those that lose their leaves in the fall) creates a browse line. Browse lines are indicative of forests with high deer densities, chronic overbrowsing, and habitat degradation. High deer densities often occur in suburban residential areas, where damage to succulent landscaping plants and gardens is common.

Browsing is rarely fatal for older trees, but stripping the bark away can be deadly if the tree is girdled (has its bark removed in a ring around the entire trunk) or if the wound introduces disease-producing organisms (pathogens). If a deer strips bark or consumes larger stems and branches, this behavior is generally a sign of malnutrition and an attempt by the deer

A browse line—a lack of forest understory vegetation—is created by prolonged browsing, including by this foraging sambar. Photo courtesy of Jackie Pringle.

Damage to ornamental landscaping, such as by this moose, often results when deer densities are high. Photo from the USFWS.

to extract limited nutrients from the cambium (the actively growing layer of the tree). Bark stripping is different from marks left by males rubbing their antlers on trees during the breeding season, and this rubbing behavior does not involve eating the bark strips. If extensive enough, however, antler scrapes can be as fatal to individual trees as bark stripping.

Large seeds have nutrients stored in them that are used by the developing plants, and these seeds are an important food source for some deer species. Plants guard these resources with physical and chemical defenses. One type of physical defense may be a thick shell, such as the seeds (nuts) of hickory (genus *Carya*) and walnut (genus *Juglans*) trees, which can prevent deer from consuming them. For other seeds, however, such as acorns (genus *Quercus*) and chestnuts (genus *Castanea*), the seed's protective shell is not enough to prevent browsing by deer. Hard-shelled nuts are collectively referred to as "hard mast," and this can be a favored food of deer and many other species of wildlife. Acorns, the well-known seeds of oak trees, contain high levels of tannins, which create a dry, puckery mouth feel. Tannins do not prevent deer from eating acorns, but they do seem to limit the amount that deer can consume in a single feeding session. There is evidence that white-tailed deer (*Odocoileus virginianus*) have salivary proteins to counteract the effects of tannins. However, field observations suggest that deer alternate between browsing and consuming mast, like acorns, during their feeding sessions.

Deer readily eat corn and other agricultural grains, and deer found in corn-producing areas tend to be in better body condition than deer without access to corn. Food plots that are provided for deer, a common management practice in North America and Europe, generally consist of cereal grains or corn in the northern latitudes, and forbs and grasses in the southern latitudes.

Soft mast includes seeds that do not have a hard shell (such as those encased in fruits) and are generally high in simple carbohydrates. Soft mast is a major summer food for most temperate-region and boreal (northern) deer species. It is readily consumed by deer, but soft mast is not edible after it has been off the plant for even a short period of time (such as windfall apples, which rot soon after they fall from the tree). A major limitation is when the fruits ripen on the plant, as only fruits within browse height are accessible to the deer.

Some deer species, such as caribou (*Rangifer tarandus*), eat moss and lichens in the autumn and winter. Sika deer (*Cervus nippon*) have been reported to feed extensively on seaweed along beaches in Japan, as have caribou along the Arctic Coast.

Acorns from red oak trees in Virginia. Photo by William McShea.

In the far north where caribou are found, forage is much more plentiful and easy to locate during the summer than the winter. Photo by Mike Boylan, USFWS.

Deer need minerals and vitamins for proper bone and antler growth. Essential minerals can be obtained by eating plants that take in minerals from the soil; by directly consuming soil that contains a high concentration of minerals, due to leaching; or by chewing on the bones and antlers of dead animals.

Deer that live in close association with people will broaden their diet. Along with rice and corn, deer also consume other agricultural crops and may cause significant damage. Various attempts to limit deer damage include hunting, guard dogs, fencing, and even noxious sprays applied to crops at critical periods (see chapter 9). Nor do deer always benefit from eating human foods. In developed countries, species such as white-tailed deer and fallow deer will eat garbage and kitchen wastes, or accept offerings of food from tourists around campgrounds. Plastic materials, which deer can consume while trying to eat human food, can accumulate in their stomachs and cause intestinal blockage and death. Anything with salt can attract deer as well, including salt used to treat icy roads. This salt preference can bring deer into a high-risk area, such as being close to vehicular traffic.

How do deer find food?

Deer are relatively selective foragers, whether they are focused on grasses, herbaceous plants, or woody buds and twigs. There is little evidence that deer can detect edible food over long distances. Instead, they use a combination of experience, group memory, and landscape shapes (topography) to position themselves in places where suitable food should be found. When they are in a potentially suitable location, deer then use their senses of smell and taste—and, to a lesser extent, vision—to locate edible food. Deer can select plants with a higher protein or mineral content, and they avoid foods containing toxic compounds. Moreover, their movement pattern changes once they encounter a suitable patch of food; their gait slows and they make more turns.

A famous ecologist said that "the world is green," and, if you are an herbivore, most habitats appear to contain an abundance of food. All this green can be deceiving, however. Forage is often limited because it is inaccessible to deer (too high up to reach), unpalatable, or indigestible. The protein and energy content of vegetation also varies considerably, according to the season. As a result, deer at higher latitudes reduce their forage intake during the winter, and thus will lose body weight until the vegetation once again "greens up" in the spring. Deer in the tropics lose weight during the dry season, when most plants become dormant, or senescent. Even when deer are surrounded by vegetation, they may be relatively selective about which plants or plant parts they will consume.

Are any deer scavengers?

Deer are not scavengers; they do not eat the flesh of dead or decaying animals. They may gnaw on bones from dead animals to gain certain minerals, but not before other creatures have removed the meat. Minerals are important to deer growth and health. Males obviously need calcium and phosphorus for their annual antler growth (see chapter 2). Mineral needs are less apparent for females, but they are just as critical during gestation and the nursing of fawns. Studies indicate that female mineral requirements during lactation can exceed those of male deer by as much as 57 percent. There is little clear evidence that deer normally lack necessary minerals, although many wildlife managers try to supplement deer diets by providing mineral blocks. Whether such mineral blocks are beneficial or not, they often do attract animals.

Because they grow large antlers, larger deer living in northern climates with poor soils may experience mineral deficiencies. Moose, for instance, have an obvious preference for mineral licks. In tropical areas, and during the summer in temperate regions, rapid water loss due to high temperatures may reduce the amount of salts in the body, which are needed for proper kidney function; white-tailed deer, elk, moose, Eld's deer, and sambar have all been reported to use mineral licks.

Forest deer, such as white-tailed deer and moose, eat dead leaves in the autumn. Fallen leaves do not retain a deer's interest for long, however, probably because nutrients are leached out by rains. White-tailed deer may consume a wide range of dead or dormant plants in winter, when other foods are not available.

How do deer digest their food?

Because of the type of cell walls in vegetation, plants are much more difficult for mammals to digest than meat. Deer, like some other hoofed mammals—such as pronghorn antelope (*Antilocapra americana*), giraffes (genus *Giraffa*), and bovids—are ruminants. They have a complex stomach, adapted for the efficient digestion of vegetation. This stomach is divided into four sections (or chambers): the rumen, the reticulum, the omasum, and the abomasum. The first three chambers, collectively called the forestomach, contain microfauna. These microorganisms—bacteria and a specialized group called ciliated protozoans—have the enzymes necessary to break down the cell walls of plants and release their nutrients.

The digestive process begins when deer chew vegetation to break it into smaller pieces. After swallowing, the food enters the rumen and the reticulum, where it is acted on by the microfauna, which break down the cellulose in the plant material into volatile fatty acids, which are an important source of fuel for the deer. A partially digested food ball, called a bolus, is then regurgitated (brought back up to the mouth) and re-chewed. "Chewing the cud" is a characteristic of all ruminants, but it occurs less often in those ruminants, such as most deer species, that consume more browse and less grass. Deer will spend a period of time foraging, and then rest and re-chew their food. However, they cannot feed continuously, as do non-ruminant animals with a simple stomach, such as elephants (*Loxodonta* and *Elephas*), giant pandas (*Ailuropoda melanoleuca*), or humans. A deer fawn or calf obtains the microbes that digest plant materials either through oral contact with its mother or from the environment. By the time these young animals are weaned, their digestive systems have an established microfauna culture (a combination of various microorganisms), so they, too, can begin feeding on vegetation.

Starches ferment in the first two chambers of the stomach, helping to digest the complex molecules within the plant materials the deer have eaten. The resulting mass (the digesta) then move into the third chamber, the omasum, where water and minerals are absorbed and the digesta are sieved. Only when the digesta pass into the abomasum—the true stomach—does final digestion occur. The abomasum is acidic and has "standard" digestive enzymes for processing proteins and fatty acids; microfauna cannot survive in that environment. Any microbes entering the abomasum are digested, becoming part of the protein component of a deer's diet.

The process from this point forward resembles digestion in non-rumi-

A sambar stag chewing browse.
Photo courtesy of M. Monirul H. Khan.

nants. The small and large intestines absorb nutrients, minerals, and water released during the passage of the digesta through the deer's four-chambered stomach. The gut-passage time of this material for deer is relatively long, lasting up to 30 hours for species such as moose. Woody plants contain greater amounts of lignin and cellulose, so if there are more woody plants in a deer's diet, the forage is less digestible, and the digestion process and gut passage takes longer. Digesta composed primarily of easily processed material, such as forbs and succulent plant parts, can have a passage time of less than 10 hours. Digestion and nutrient absorption also occur in the cecum, located between the small and large intestines. The cecum is a relatively small organ in deer species, but is important for nutrient absorption during stressful periods.

Do deer store food?

Deer do not store (or cache) food, as do certain rodents, shrews (family Soricidae), and other mammals. As with most mammals, however, deer store food energy as fat, or adipose tissue. Adipose is a connective tissue that is a compact means of storing energy for future use. Like all animals, deer use these fat stores when their intake of energy does not match the amount they need to use for physical activity, growth, reproduction, the maintenance of body temperature, and other requirements.

Muscle cells can also be transformed and used for energy, but drawing upon these cells is only a short-term solution, one that reduces the animal's body condition. During prolonged periods of stress, however, once the fat stores are used up, muscle tissue can be broken down (catabolized) to produce energy. Starvation in deer is preceded by a loss of muscle tissue and a general wasted appearance in the animal.

In deer, fat stores are first depleted in subcutaneous areas (under the skin), such as the rump; then around internal organs (the viscera), including the kidneys; and, finally, in the bone marrow. The rump is the dominant fat storage area for deer. In larger deer, a hollowing of the pelvic area indicates low body fat. For males, a loss of fat around the neck and chest is also indicative of low fat stores.

Once fat depletion has extended to the viscera, a traditional technique for measuring body condition in deer has been an index based on the amount of visible fat around the kidneys. Similarly, testing the marrow within the large leg bones of deer provides another way to estimate the degree of fat loss. Loss of marrow fat only occurs when a deer has exhausted most of its other body fat and is near starvation. Both the kidney and bone marrow tests can only be conducted postmortem (after death), but ultrasound has been used with some success to measure the rump fat in live elk and white-tailed deer. These various measurements are comparable to a calculation of total body fat, which is obtained by processing the entire carcass of the deer. A rough measure of body condition for captured deer is the animal's body weight relative to standard skeletal measurements, such as chest girth, lower jaw length, or hind leg length.

For species living in temperate regions, fat is primarily deposited during the summer and fall. If an adequate quantity and quality of forage is available, males will consume and digest food throughout most waking hours during the summer and early autumn. Adult females are nursing their young early in the summer, so they cannot deposit significant fat stores until the fawns or calves are weaned. Females begin intensive feeding later into the fall, at a time when males are engaged in establishing breeding territories and preparing for the rut. For deer species living in temperate-zone forests, the production of fruits and mast is the last major pulse of food for them before the onset of winter. In oak-dominated ecosystems, mast production in the fall is a significant determinant of a deer's body-fat condition when entering the winter season. For more open-habitat species of deer, grass and forb production during the summer is more critical. Agricultural crops supplement the diet of deer species in many regions. The timing of when these crops mature can greatly affect the survival, movements, and behaviors of deer. Deer species in temperate regions have also been found to undergo lipogenesis in the fall, a process where fat deposits are formed regardless of overall food consumption. Therefore, under either good or poor autumn forage conditions, both adult and young deer will form and deposit fat.

The total amount of fat a deer stores is a function of its body size. Larger deer can carry more fat. They also have a lower surface-to-volume ratio, so they lose less body heat and expend relatively less energy during the winter. None of the smallest deer species exist in extreme northern or southern latitudes, with roe deer being the smallest of the temperate-region species. For all temperate species, however, fawns and calves are most at risk during the winter. Adult deer deplete fat and muscle tissue in winter, but they can lose about 30 percent of their body weight before they die of starvation. It is much more common for young animals, rather than the adults, to starve during this season. Their energy intake may go into fat deposition, but their small body size and low total fat stores make them vulnerable to severe low temperatures. Late-season birthing (due to late mating), or the low nutritional status of the mother (especially when combined with a multiple birth), can limit fawn size by the onset of winter and thus reduce the chances of survival for these young.

Tropical deer generally do not store as much fat as species living in temperate areas—nor do they need to. In general, the amount of body fat stored by temperate-zone deer in the fall is a strong determinant of the probability of their overwinter survival. For tropical deer in monsoon environments, fat probably enables them to survive a lack of food during extended droughts; they can also catabolize fat tissues, releasing metabolic water when environmental water is scarce.

Chapter 8

Deer and Humans

Do deer make good pets?

Deer are wild animals, not pets. Grown deer can be dangerous and unpredictable, especially males during the rut (mating season), even if they have become fairly accustomed to their surroundings. Without permits, keeping a wild animal is illegal almost everywhere in the United States; most city and county ordinances also restrict keeping wild animals as pets.

Should people feed deer?

Many well-intentioned people feed deer in their backyards, especially in northern areas where deep snow, ice, and freezing temperatures prevail during the winter. In the United States, a few state and federal wildlife management agencies in northern regions operate larger-scale supplemental feeding programs as needed during particularly harsh winters, especially for elk (*Cervus elaphus*). Feeding deer is a controversial, highly debated issue, however.

When people feed deer, they can do more harm than good—to individuals as well as to the local population—in part by using the wrong foods. It takes deer several weeks to become used to, and be able to digest, a high-carbohydrate diet of grains, alfalfa, and corn in the winter. Deer may have a stomach full of food but still starve, because their bodies can't effectively process this material. This results in too many acids in their body, a condition known as acidosis.

Several other problems may result from people feeding deer. As deer become artificially concentrated around the feeding site, they are more susceptible to the spread of parasites and diseases (including chronic wasting disease and brucellosis; see chapters 5 and 10). Large crowds of deer in one place may also damage their existing natural habitat, as well as ornamental plants and landscaping. If feeders are located far from natural bedding areas, deer will expend valuable energy in moving to them. If deer cross roads to get to feeders, they become more vulnerable to collisions with cars and trucks. Larger, dominant adults may keep younger and weaker deer away from the feeders. As a result, these excluded deer, often fawns, will have used critical energy stores getting to supplemental food for a minimal return in energy gained. Deer can quickly adapt and become dependent on supplemental food, so people are obligated to continue feeding them until natural forage is available in the spring. Providing appropriate food, in sufficient amounts, to large numbers of deer for several months can be an expensive proposition. Moreover, in many states there are laws against people feeding deer.

Opponents of large-scale supplemental feeding programs by wildlife management agencies in North America and Europe use many of these same arguments. However, when agencies do provide supplemental food to deer, they usually start early in the season, to allow time for a deer's digestive system to adjust. Also, supplemental feeding is often good for public relations, because people do not like to see news coverage of starving deer, even if the overall population is not threatened. Winter feeding can be justified when a deer population is small and well below the carrying capacity of the habitat (the ability of the land to support the deer living on it)—although the natural environment should normally provide sufficient food when the numbers of deer are low. Another reason is that supplemental feeding may keep deer away from private lands and domestic livestock.

Other circumstances that may justify winter feeding include the loss of traditional winter range because of drought, wildfire, fencing, or development. For example, the National Elk Refuge in Jackson Hole, Wyoming, feeds about 6,000 elk each winter. Likewise, the state of Wyoming feeds about 13,000 elk at 22 sites annually. Large-scale supplemental feeding programs that maintain populations at densities beyond the carrying capacity of the habitat, however, are not recommended. Deer at high densities, maintained by supplemental feeding, will damage natural vegetation at "unnatural" rates.

The best approach to ensure the survival of deer herds during severe winters—and to avoid supplemental feeding and its associated problems—is to maintain population densities within the carrying capacity of their range. It is also important that deer have access to traditional migration routes and pastures, and that the quantity and quality of available food during the spring, summer, and fall allow deer to enter the winter in good physi-

It is easier for deer in good physical condition, such as this moose, to survive harsh winter weather. Photo by Ronald Laubenstein, USFWS.

cal condition. Finally, good winter cover in the form of conifers—evergreen trees such as hemlock (genus *Tsuga*), spruce (genus *Picea*), and fir (genus *Abies*)—is essential as well. Conifers help reduce the amount of snow reaching and covering the ground, so moving is easier and less energetically costly for the deer. Evergreen trees also shelter deer from the wind.

Do deer feel pain?

Yes, deer feel pain—as does any animal that has a developed nervous system. Pain receptors, which are part of the nervous system, are highly adaptive, adjusting to various situations. They alert deer to dangerous and potentially harmful stimuli, and they help individual animals avoid such situations. Experts don't know if the pain deer feel is exactly the same as in humans, however.

What should I do if I find an injured deer?

Handling deer can be very dangerous. The best course is to call a licensed animal rehabilitation facility, the local Humane Society, an animal control officer, or your area's Department of Natural Resources office. Wildlife rehabilitators are trained to provide care for injured, sick, or orphaned animals, including deer, with the goal of returning them to the wild. Contact information for a local rehabilitator can be obtained at www.wildliferehabber.org or related sites. Although most people want to help an injured animal, it is important to realize that, from a population standpoint, the fate of one individual will not affect the well-being of a local deer herd.

What should I do if I find a fawn?

If the fawn appears to be healthy, leave it alone and quietly walk away. Young deer are normally left alone while their mother feeds, and she is often close by. Handling the fawn will stress both animals. If the fawn is sick or hurt, or if you know that the mother has died, you can call a licensed wildlife rehabilitator to come get the fawn. If you feel it is imperative to move the fawn yourself, cover it with a cloth, and wear gloves as you carefully pick it up. Put the fawn in a box or animal carrier with a cloth on the bottom, and keep it in a warm, dark, quiet location, away from children or family pets. Bring it to a wildlife rehabilitator as soon as possible. Do not attempt to feed the fawn or handle it. Again, it is illegal to keep a fawn, regardless of whether you intend to release it into the wild or not—and usually the best policy is "hands off."

What is the best way to observe deer?

Deer are usually most active at dawn or dusk, so this is the best time to look at them. The best way to observe them depends on the species and their preferred habitats. Many larger species, including white-tailed deer (*Odocoileus virginianus*), mule deer (*O. hemionus*), fallow deer (*Dama dama*), and elk or red deer (both *Cervus elaphus*), usually are common and readily visible as they feed in open areas. It is far less likely that you would casually encounter a threatened or endangered species (see chapter 10). Endangered species of deer are usually few in number and live in restricted or remote geographic locations. Regardless of how common they are, small, solitary, cryptic (well-camouflaged) species residing in dense cover—such as tufted deer (*Elaphodus cephalophus*), brocket deer (genus *Mazama*), muntjac (genus *Muntiacus*), or pudu (genus *Pudu*)—may require a pair of binoculars and many hours of patient waiting to see them.

Although they may appear to be abandoned, deer fawns should be left alone. Photo by W. J. Berg, USFWS.

Larger species, including these fallow deer hinds, are often easily seen. Photo courtesy of Jackie Pringle.

Smaller species of deer, such as this southern brocket, are often hidden in dense vegetation. Photo courtesy of José Mauricio Barbanti Duarte.

One reward of looking at deer for any length of time is that you will begin to recognize individuals by their size, distinctive marks, or color patterns. Even inexperienced observers can usually tell adult males and females apart during most of the year, and distinguish adults from young animals. Different behaviors—alert signals, feeding, grooming, play, aggression, and others—also become apparent when studying deer in a group. You will also notice that some behaviors, including reproductive activities, vary seasonally.

How economically important are deer today?

Deer are among the large mammals that have long provided humans with the necessities of life—meat, hides for shelter and clothing, bones and antlers for tools and artwork—from early Paleolithic peoples to present-day indigenous (native) societies and subsistence hunters (those who hunt for survival, not sport). Throughout the world, deer continue to fulfill both positive and negative economic roles.

In the late 1800s in the United States, deer were an integral part of what is known as market hunting. With few regulations on hunting seasons or bag limits, uncontrolled deer hunting—in combination with the loss of forests and other habitat—dramatically reduced deer numbers. Meat from market hunting was used to supply mining, military, railroad, and lumber camps, as well as people in growing urban and rural population centers. Aggressive restoration and conservation efforts eventually allowed species to recover to the point that sport hunting for deer currently is a multi-billion-dollar industry annually. Although the figures vary each year, in the United States alone there are about 11 million deer hunters who are in the field an average of 12 days each season. According to data from the International Association of Fish and Wildlife Agencies, and from surveys published by the U.S. Fish and Wildlife Service, deer hunting annually generates in the neighborhood of $25 *billion* in retail sales for equipment, travel, food and lodging, and a variety of associated activities. The economic value of deer from non-consumptive users like hikers, photographers, and others also amounts to considerable sums annually. Deer hunters and non-consumptive users in Europe also spend a significant quantity of money each year. In South America, sport hunting is important, but poorly regulated. Deer hunting in Asia is largely unregulated.

Besides hunting deer since Paleolithic times, people have been herding reindeer for thousands of years. In terms of their scientific classification (i.e., taxonomically), reindeer are the same as caribou (*Rangifer tarandus*), but they are somewhat domesticated. Unlike truly domesticated mammals, breeding in these herded animals (sometimes referred to as "exploited captives") generally remains natural rather than artificially controlled by ranchers. The ability of reindeer to survive in the extremely harsh conditions of northern Scandinavia and Russia provided a livelihood and natural resources for the people who herded them. As such, reindeer have been used for food, shelter, and transportation for millennia. The economic importance of deer also reaches back thousands of years in another way: in the use of their antlers and other body parts in traditional Asian medicine, a practice that continues today. Domestication of other species of deer is a

Uncontrolled market hunting at the turn of the nineteenth century dramatically reduced almost all big-game species, including these white-tailed deer, throughout much of the United States. Photo from the Illinois Department of Natural Resources.

Traveling by reindeer in Russia, circa 1900. Photo from the Library of Congress.

relatively new initiative in many parts of the world, where they are raised on game farms for a variety of products.

What are deer game farms and game ranches?

Game farming is the breeding and raising (husbandry) of wild animals in penned conditions. Game ranching is the management of free-ranging wildlife, usually on private property, with or without fences. Deer farming occurs throughout Europe, Scandinavia, and Asia, as well as the United States and Canada. New Zealand leads the world in the number of farm-raised deer, mainly red deer. Additional species—sika deer (*Cervus nippon*), North American elk (*C. elaphus*), fallow deer, white-tailed deer, and sambar (*Rusa unicolor*)—live on deer farms in dozens of different countries worldwide.

Almost all U.S. states have deer farms, as do most Canadian provinces.

However, the commercialization of exotic (non-native) as well as native deer on game farms or ranches remains a relatively new phenomenon in North America. The North American Elk Breeders Association, the Exotic Wildlife Association, and the North American Deer Farmers Association (NADeFA) are among the various organizations that have been established to promote the industry, set standards, and work with government agencies to develop regulations. According to NADeFA, more than 15,000 deer farms/ranches operate in the United States, with an annual production worth about $3 billion. Nonetheless, the commercial importance of the deer industry in North America remains a small fraction of that of the domestic cattle industry. Deer farms in North America are closely regulated to prevent the spread of disease from captive to free-ranging animals, especially for chronic wasting disease (see chapters 5 and 10).

How and where did deer farms begin?

Deer farms began in New Zealand, which is not what one would expect, because deer are not native to that country. The many new species of deer initially brought into New Zealand and released in the wild quickly began to overpopulate certain areas of the country and destroy habitats. As the disruption of the forest community became increasingly obvious, people attempted to reduce (cull) deer populations by shooting the animals, often from aircraft because of the inaccessible terrain. It soon became apparent that it would be far more profitable to market the meat of all these culled deer instead of leaving the carcasses on the ground. So, during the 1960s,

Traditional Chinese medicine uses almost all parts of a deer, including the penis (*left*) and tendons (*right*).
Photos by Chris73.

hunters killed tens of thousands of deer, then removed them from the rugged backcountry by helicopter so as to process and export the meat (venison). The operation proved so efficient, however, that eventually too few deer remained to sustain the enterprise. At this point, it became clear that, instead of shooting free-ranging deer, it would be much better to capture them alive, confine them in pastures, and raise them like cattle or other livestock. Thus, game farming began in New Zealand in the 1970s and continues to flourish. The country has more than 4,000 game farms that raise over 2 million deer and export literally tons of venison, along with antler velvet, hides, and other deer products, some of which are used in traditional Asian medicine. Game farms and ranches for deer are continuing to expand in numerous other countries as well.

What parts of deer are used in traditional Asian medicine?

Traditional Asian medicine primarily uses hard antlers and the velvet from growing antlers. Numerous Chinese medical treatises from 168 B.C. to the present day—referred to collectively as *pênts'ao*—mention 28 different parts of a deer's body that they claim are able to cure a variety of ailments and debilities. Almost all the organs, the teeth, blood, semen, fat, the penis, tendons, and other body parts, even excrement, are believed to provide beneficial effects. Today's practitioners often blend ground-up hard antler, velvet, or other body parts with medicinal plants to sell as over-the-counter pills, extracts, or tonics.

Chapter 9

Deer Problems (from a human viewpoint)

Can deer be pests?

Too much of anything can cause problems, and too many deer are no exception. Overabundant deer can create difficulties in several ways. Of course, what constitutes "too many deer" is a value judgment that certainly will vary among individuals. Biologists consider deer to be overabundant when they disrupt people's lives, when they negatively affect other animal species, when they have harmful consequences for their own populations, or when they disrupt the structure and function of plant communities. Excessive deer populations can transmit certain diseases to humans, damage crops, slow the rate of forest regeneration, or result in more collisions with vehicles.

The economic aspects of deer-vehicle collisions are not trivial. At an average vehicle-repair cost of $2,800, auto-deer collisions amount to $4.2 *billion* each year in the United States, not to mention the terrible waste of a wildlife resource. There is a significant social cost as well—150 to 200 people die annually in the United States from such collisions. In other countries throughout the world, millions of deer are also hit by vehicles. These problems involve social and political factors, as well as biological ones—and they are an increasingly important issue for wildlife managers and conservationists as they work to achieve balanced, multiple-use deer management goals.

Unfortunately, dead deer on roadways are a common sight throughout much of the world. Photo from the Idaho Department of Transportation.

How do I reduce deer damage?

Deer damage to property in rural, suburban, and urban areas is becoming a significant problem. When fields or gardens are large, or when deer populations reach high densities, it becomes increasingly difficult to protect vegetation. Numerous websites are devoted to ways to reduce the damage by deer, along with a multitude of books and publications from government agencies and universities with such titles as *Outwitting Deer*; *Solving Deer Problems—How to Keep Them Out of the Garden, Avoid Them on the Road, and Deal with Them Everywhere!*; and *Deer-Proofing Your Yard & Garden*. These publications, and many others, often focus on species of trees, shrubs, vines, ground cover, and other types of vegetation that are relatively resistant to deer browsing, due to an unpleasant taste, a strong odor, thorns, or similar characteristics. Deer browsing is only one aspect of the problem, however. Naturalists and wildlife specialists are also investigating a variety of deterrents to keep deer out of agricultural areas, residential landscaping and gardens, and highway rights-of-way.

There are several management options to reduce deer damage. Four are related to the landscape: (1) exclusion; (2) repellents; (3) silviculture; and (4) diversionary feeding. Other options are geared toward the deer themselves: (1) live capture, removal, and release; (2) fertility control; (3) culling by sharpshooters; and (4) public hunting. These different ways of reducing deer damage involve tradeoffs, because there are positive and negative aspects to all of them. The "best" method is the one that is least costly and most effective in limiting the amount of damage for the longest period of time. Methods also depend on site-specific characteristics and circumstances. Equally critical is what is acceptable to people in terms of their values and concerns.

Landscape Options. *Exclusion* methods, intended to keep deer out of certain areas, include a variety of approaches: different configurations of wire or electrical fences surrounding fields or gardens, wire or plastic protective cylinders around individual plants or seedlings, and netting or boards piled up to protect haystacks. Deer can also be frightened away from fields by occasional loud noises or the use of startle devices, such as propane-powered "cannons" fired at set intervals, guns, or starter pistols; or by the use of guard dogs. However, fencing becomes prohibitively expensive for enclosing large tracts of land, and deer usually get used to noises from scare devices fairly soon and then ignore them. Guard dogs cannot always be on site, and they can only patrol limited areas.

There are many *repellents* available that smell or taste bad to deer, including simple remedies like hanging up pieces of bar soap or bags full of human or dog hair. A variety of commercial products contain various chemical formulations, including putrefied (rotten) meat or eggs, ammonia, capsaicin (the active ingredient in hot peppers), and other compounds. Repellents are of limited use over large expanses of land, however, and they must be applied repeatedly.

In forested areas, *silvicultural* (forest management) practices include various patterns, or regimes, of planting or cutting trees that can affect deer. For example, clearcutting (removing all trees) on several acres will result in a dramatic increase of new vegetation, such as grasses and forbs, available to deer during the next few years. This may temporarily divert deer away from croplands or other protected sites, but, in the long run, this method only produces more deer, because it provides more food for them to eat.

Diversionary feeding is a similar practice. Planting food plots of grasses or forbs adjacent to agricultural fields and pastures can draw deer away from more sensitive areas. Not only can this be expensive, however, but, like clearcutting, it eventually may cause deer populations to increase.

Deer Options. To reduce local deer populations, the public in suburban areas usually favors non-lethal means, such as *live capture and removal*, with subsequent release elsewhere. Live capture of deer in box traps or by chemical immobilization (delivered with tranquilizer darts) is fairly easy, although it usually costs several hundred dollars per animal. It is much more problematic, however, to find a place to relocate deer where they will not continue to cause damage. Survival of relocated deer is also often very low. Finally, because of concerns about spreading disease, in most states deer relocation is now illegal.

Likewise, the public often favors *fertility control*, using chemosterilants (chemical compounds that interfere with fertility) on female deer. The method may be promising in the future, but currently its applicability is fairly limited. Besides the expense, injecting or implanting these reproductive inhibitors is time and labor intensive, and they only work for a year or two. There is, however, a new, approved, single-dose contraceptive for deer (trademarked as GonaCon) that lasts for several years. Nonetheless, to effectively control fertility, wildlife managers must treat a large proportion of the deer population, always a challenging proposition. There is also the issue of the treated deer being consumed by other wildlife or by people, and what effect the chemicals have on deer meat (venison).

In certain situations, professionals can *cull* (shoot) excessive or problem deer in city parks or suburban neighborhoods. Shooting usually occurs from vehicles or elevated stands. Culling can be effective, but openly killing large numbers of deer can cause public concern, as well as vehement opposition by animal rights groups.

Most wildlife managers agree that *public hunting* is the best, most cost-effective approach to keeping local or regional deer herds at acceptable population levels. However, hunting may be problematic in urban/suburban settings; ultimately, each community must weigh issues of cost, safety, environmental protection, and welfare of the deer.

How many deer are hit by vehicles?

> There are few more dramatic manifestations of human-wildlife conflict than squealing brakes, a sickening crunch, flying gravel, and then silence except for the weakly spasmodic scrabbling of a semi-pulverized deer as it lies dying on the side of a highway.
>
> J. T. du Toit, *Human-Wildlife Conflicts*

It is not surprising that the number of deer-vehicle collisions (roadkills) has increased throughout the world as more highways are built and both traffic volume and speeds increase. In North America and Europe, moreover, deer and human populations have both been growing.

In the United States alone, there are an estimated 1.5 million deer-vehicle collisions a year—primarily involving white-tailed deer (*Odocoileus virginianus*), but also mule deer (*O. hemionus*), elk (*Cervus elaphus*), and moose (*Alces alces*). Although 1.5 million road-killed deer is a lot, it is still a conservative estimate. The actual number of deer hit each year is certainly higher, because many such incidents are not reported. Of course, deer-vehicle collisions are not unique to the United States; many hundreds of thousands occur annually throughout the developed world. Drivers hit an estimated 500,000 deer annually in Europe, resulting in $1 billion in property damage and about 300 human fatalities. Germany accounts for about 140,000 road-killed deer a year. More than 10,000 moose and 50,000 roe deer (*Capreolus capreolus*) are struck in Sweden each year. In Great Britain, there are an estimated 74,000 deer-vehicle collisions annually—primarily with roe deer, fallow deer (*Dama dama*), and red deer (*Cervus elaphus*), as well as with an introduced species, Reeve's muntjac (*Muntiacus reevesi*). In Japan, automobiles are hitting increasing numbers of sika deer (*Cervus nippon*).

In North America and Europe, most deer-vehicle accidents occur in late autumn, during the mating season, when deer are most active and least cautious. A second peak takes place in early spring, with the green-up of vegetation along the sides of roads and highways, as deer move to these areas to forage on the new growth. As might be expected, drivers strike most deer between dusk and dawn, when the animals move about the most and driver visibility is poorest. Roads that cross migratory routes of deer increase the problem.

Given the economic impact and loss of life (for both humans and deer), it is not surprising that many federal, state, and local agencies work to reduce the number of deer-vehicle collisions in the United States. The U.S. Department of Transportation and the Federal Highway Administration are involved in mitigation efforts (attempts to moderate or reduce the problem) on highways, as are many state transportation agencies. Groups such as the Insurance Institute for Highway Safety, the Insurance Information Institute, and a number of universities are studying the problem as well. The University of Wisconsin houses the Deer-Vehicle Crash Information Clearinghouse, for example, and the Center for Transportation and Environment is located at North Carolina State University. Utah State University and Mississippi State University house branches of the Berryman Institute, founded to reduce problems between wildlife and people through research, education, and outreach. In Great Britain, the Highways Agency and the Scottish Executive both fund the collection of data and work to reduce collisions through the efforts of the Deer Initiative, established in 2003.

Where are deer most often hit?

About 50 percent of all deer-vehicle collisions in the United States each year occur in only 10 states. Pennsylvania or Michigan often lead the nation in the number of road-killed deer. Drivers in Pennsylvania usually hit more deer on roads each year than hunters in neighboring Maryland legally harvest. In some small northeastern states, more deer may be hit on their highways than are legally taken each year during the hunting season. Because Hawaii has no native species of deer, you are least likely to hit a

White-lipped deer, or Thorold's deer (*Przewalskium albirostris*). Photo courtesy of Jiagong Zhala.

Female red brocket deer (*Mazama americana*). Photo courtesy of José Mauricio Barbanti Duarte.

Fallow deer (*Dama dama*) females.

Photo courtesy of Jackie Pringle.

Red deer (*Cervus elaphus*).

Photo courtesy of Rory Putman.

Sambar (*Rusa unicolor*). Photo courtesy of Jiagong Zhala.

Persian fallow deer (*Dama mesopotamica*). Photo courtesy of Eyal Bartov, Israel.

Barasingha (*Rucervus duvaucelii*) females. Photo courtesy of Kaleem Ahmed.

Chinese water deer (*Hydropotes inermis*). Photo courtesy of Chen Min.

Caribou (*Rangifer tarandus*). Photo courtesy of Seth Stapleton.

Female European roe deer (*Capreolus capreolus*). Photo courtesy of Jackie Pringle.

Southern pudu (*Pudu puda*). Photo courtesy of JoAnne Smith and Werner Flueck.

Brown brocket deer (*Mazama gouazoubira*). Photo courtesy of José Mauricio Barbanti Duarte.

Male white-tailed deer lock antlers and spar during the breeding season to establish dominance and access to females.

Photo courtesy of David Tuttle.

A young white-tailed deer fawn nurses. Photo courtesy of David Tuttle.

A "reunion" of an adult female white-tailed deer with her fawn, and a yearling male from the previous year with his spike antlers in velvet. Photo courtesy of David Tuttle.

Close-up of a male white-tailed deer with polished antlers. Photo courtesy of David Tuttle.

Roe deer, such as the one browsing in the photo, are commonly involved in collisions with vehicles in England. Photo courtesy of Rory Putman.

deer there. Unfortunately, however, two species have been introduced, so deer-vehicle accidents do occur even there.

What are ways to reduce deer-vehicle collisions?

Reducing the number of vehicles or the number of deer should lower the risk of collisions. Because reducing the number of vehicles is unlikely, mitigation is directed instead toward modifying the behavior of motorists to reduce their speed and become more aware of deer in high-risk areas. Another option is to enhance vehicle technology, so drivers can more readily avoid collisions with deer. Wildlife managers have also attempted to reduce deer activity around roads, modify the roadside environment, or make roads less accessible to deer. Despite these many different approaches, none have been especially effective.

There are four main methods to prevent collisions involving vehicles and drivers. First are *deer crossing signs*. Diamond-shaped signs warning of deer crossing are familiar to everyone who drives on major roadways. These signs are often associated with reductions in the posted speed limits. Nonetheless, numerous studies have shown that drivers rarely decrease their speed because of the signs, and that the signs are ineffective in lowering the number of deer collisions. Deer-crossing signs may be so prevalent that drivers simply ignore them. As a result, more elaborate efforts have been made to enhance the basic signs with lighted messages or lighted symbols of a jumping deer. Some lighted signs even have a microwave or infrared sensor that is activated by deer coming into the area; the sign then flashes a warning message to motorists.

The second is *vehicle technology*. Engineers have tested infrared devices

Common warning signs alongside the highway—like these for deer (*left*), moose (*middle*), and caribou (*right*)—have little effect on reducing deer-vehicle collisions. Photos by George Feldhamer (*left, middle*) and from the Swedish Transport Administration (*right*).

that allow a driver to better see or otherwise locate deer that are on the road at night. These devices display a thermal image of a deer (activated by the deer's body heat) much sooner than a driver would see one with normal headlights. The devices are quite expensive, however, and available in only a limited number of vehicles.

Third is *highway lighting*. A driver's ability to see is lowest at night, the time when deer are most active and when most collisions occur. Increased roadway lighting would seem an obvious solution. Nonetheless, although the area under the lights is brighter, the shadows outside of that area—where deer are likely to be—appear deeper, potentially making it more difficult to spot them. Scientific testing done so far suggests that highway lighting has little, if any, effect in reducing collisions.

The fourth method is the use of *deer whistles*. These simple, very inexpensive, bumper-mounted devices have been used since the 1970s and are popular with many motorists. When a vehicle reaches 30 mph (48 km/hr), the whistles are supposed to create an ultrasonic noise that deer hear and avoid. The concept is simple and appealing. Scientific evidence indicates, however, that deer do not react to the whistles—probably because they do not hear them—and that whistles are ineffective in reducing collisions.

There are also six common methods of changing deer behavior. The first is *intercept feeding*. This involves planting food that diverts deer away from roads and reduces their incentive to be near roadways. This method, like others, remains inconclusive, and it certainly is dependent on local herds, seasonal habitat, and patterns of property ownership.

Second is *vegetation management*. The goal is to remove palatable vegetation that attracts deer, plant vegetation that deer find undesirable, and mow the right-of-way on a schedule that attracts fewer animals while not compromising aesthetics or the safety of motorists. Unfortunately, the resistance of various plants to deer browsing is highly variable, and if deer are hungry enough they will eat a wide array of plants.

Next are *deicing alternatives*. Many states and provinces use salt in the winter to reduce snow and ice on the roads. In the spring, much of this salt finds its way into roadside pools and onto vegetation—essentially producing an artificial salt lick that may attract deer. Officials have proposed alternatives to salt, such as calcium chloride, and recommended draining roadside pools. There has been very limited research in this area, however, and a direct relationship between roadside salt and deer-vehicle collisions remains inconclusive.

A fourth method is the use of *repellents*. Scientific journals and popular magazines have reported on an extensive array of chemical and biological repellents to keep deer away from vegetation. Researchers have evaluated numerous potential repellents on whitetails, mule deer, caribou (*Rangifer tarandus*), and elk in the United States, and on other species of deer in Europe and elsewhere. Of the hundreds of repellents tested, putrescent (rotting) egg and natural-predator odors have proven the most effective at keeping deer away from roads. A downside to repellents is that they must be applied every few weeks, and their overall effectiveness remains questionable, because deer can simply move to nearby sections of the road where there are no repellents.

Fifth is *fencing*. Wire fences 9 feet (3 meters) high have been used to keep deer away from highways for many years, and numerous investigators have studied their effectiveness. Although expensive, 9-foot-high fencing, properly installed and maintained, can reduce the number of deer on roadways. Whether this translates into fewer roadkills is not always clear. Electric fencing can also reduce the number of deer on the right-of-way. A variety of fences have been tested that also have escape routes if deer end up on the wrong side of the fence. These include one-way gates, and underpasses or overpasses that allow deer either to get to the opposite side without crossing the road or to escape from the roadway if they manage to get onto it.

Last is the use of *reflectors and mirrors*. These devices are placed along roadsides to reflect the headlights of oncoming vehicles off to the side of the road. The objective is to freeze or distract deer that are about to enter the roadway, at least until the vehicle is past. Obviously these devices are only useful at night, although that is when most collisions with deer occur. Despite having been used for decades, there is no clear general agreement as to the effectiveness of reflectors and mirrors in decreasing vehicular collisions with deer.

Many of these approaches may hold promise in specific regions and landscapes, especially when several methods are combined, but conclusive results are often difficult to achieve. Regardless of any of the other methods, one of the best approaches to reducing deer-vehicle collisions—as well as damage to forests, crops, and landscaping—is to keep local population densities within reasonable and manageable limits.

Do deer have diseases that are contagious?

Deer normally do not transmit (directly pass on) diseases to humans, but they can serve as hosts to various parasites, such as ticks, that do affect us. These parasites, or vectors, carry what are called etiological agents, viruses or bacteria that actually cause the disease. Diseases that non-human mammals transmit to people are collectively referred to as zoonoses. A prime example of a deer zoonose is Lyme disease. Although first identified in Europe and Asia, the disease is named for Lyme, Connecticut, where it was first recognized in the United States in 1977. The etiological agent is the bacterium *Borrelia burgdorferi*. The primary hosts for Lyme disease in central and eastern North America are white-footed mice (*Peromyscus leucopus*) and white-tailed deer. Confirmed cases in America have increased from about 1,000 people in 1982 to more than 27,000 as of 2007, associated in part with larger deer populations.

Another well-known, tick-borne disease carried by deer is Rocky Mountain spotted fever (RMSF), caused by the bacterium *Rickettsia rickettsii*. The name of this disease is misleading, as it occurs from western Canada, south through the United States and Central America, to Brazil. Deer are only one of many other species of mammals that serve as hosts for ticks with RMSF. Likewise, deer can carry and become infected with the agent that causes bovine tuberculosis, *Mycobacterium bovis*. Other tick-borne diseases possibly carried by deer include southern tick-associated rash illness, with symptoms very similar to Lyme disease, as well as ehrlichiosis, an infection that kills white blood cells.

A disease of recent concern in North American deer, elk, and moose is chronic wasting disease (CWD)—one of a group of transmissible, progressive, neurodegenerative diseases (affecting the body's nervous system) in mammals (also see chapter 5). The infectious agents in these diseases are called prions, which are small, modified forms of cellular protein. A prion disease (technically, a spongiform encephalopathy) causes large open areas to form in parts of the brain, producing a loss of motor control, paralysis, dementia, and eventually death. One such prion disease is bovine spongiform encephalopathy, or mad cow disease. Known prion diseases in humans include kuru, Creutzfeldt-Jakob disease, and Gerstmann-Sträussler-

Scheinker syndrome. In deer, CWD causes weight loss, excessive salivation and urination, and tremors. Infected animals probably die within a few months after coming down with the disease. Currently, there is no evidence that CWD is transmitted from deer to humans, but this remains a very active area of research.

Can deer negatively affect forest regeneration and structure?

The quality and quantity of food and shelter that forests provide have a significant impact on deer populations. On the other hand, deer can also have major impacts on forests. Constantly high densities of deer can harm forests and other habitats. With too many deer, their browsing can keep many young trees from regenerating and reduce the numbers and variety of herbaceous plants. This decreased structural diversity of the forest severely reduces its overall biodiversity. Throughout much of eastern North America, for example, increasing populations of whitetails have been a significant factor in the shift of forest communities from predominantly oak (genus *Quercus*) and hickory (genus *Carya*) trees to maple (genus *Acer*) and beech (genus *Fagus*). In the western United States, high densities of deer and elk are thought to have a negative impact on stands of aspen (genus *Populus*).

How can exotic species of deer cause problems?

> Man is an inveterate and incorrigible meddler, never content to leave anything as he finds it but always seeking to alter and—as he sees it—to improve. . . . One of the ways in which man has sought to modify the natural environment is by the introduction of animals (principally mammals and birds) and plants throughout the world.
>
> C. Lever, *Naturalized Mammals of the World*

Serious problems have resulted from non-native, or exotic, species of deer that have been transported throughout the world. Exotic species—also called introduced, alien, or non-indigenous species—are those taken to areas outside of their native geographic range. Most exotics were transported unintentionally, such as small, or cryptic (not easily seen), mammalian species like the house mouse (*Mus musculus*), which was an uninvited stowaway on sailing ships to the New World. Deer are certainly not small and cryptic, however. They were intentionally introduced into numerous countries around the world.

Managing wildlife for commercial purposes is a controversial issue, and it has generated debate among and within the public, regulatory and wildlife agencies, and the scientific community. There are many perceived positives to introduced deer. Wildlife ranchers, for example, commonly stock exotics like sika deer, fallow deer, and red deer, because they are easy to work with. Exotics provide a diversified resource for the production of venison, antlers, antler velvet, and breeding stock—not to mention a trophy animal for hunters without the expense of overseas travel. Transfers of deer from one locale to another can help ensure the survival of threatened or endangered species. And, more generally, people are attracted to deer and value the aesthetics and other non-consumptive benefits of these animals.

There are equally valid arguments for opposing the introductions of non-native species of deer. One is a political concern: the individual private-property rights of citizens attempting to develop a new business versus the traditional view that wildlife is not livestock and therefore belongs to the public at large, with its oversight entrusted to governmental regulatory agencies. Other issues relate to the ecological impact of non-native species. Introducing exotics into the habitats of native species often fails, because the exotics have not evolved alongside the area's native plants and animals. The introduced species simply cannot adapt to what, for them, is very different vegetation, climate, competition, and a mosaic of other interrelated factors. It is more problematic, however, when an introduction is *too* successful. The exotic species may increase dramatically in their overall numbers and spread far beyond the initial release site, because normal ecological balances do not keep the population under control. Exotic deer populations, unchecked by native predators, climate, or other factors, may then cause severe damage to the ecological integrity of native plants and animals. Such exotics are often referred to as invasive species.

A prime example of invasive deer species occurred in New Zealand. The country has no native deer, and some of the eight exotic species of deer imported there overpopulated, destroying the native vegetation. Not only was the habitat no longer able to support the introduced species of deer, but other types of native animals suffered from food shortages, increased erosion, and associated environmental problems. The same issues can arise in agricultural areas, where the combination of native species and invasive exotics can severely increase crop damage.

In addition, exotic deer can often outcompete native species. Sika deer in North America have displaced white-tailed deer in areas where the two species overlap. Even though they are much smaller than whitetails, sika deer are highly aggressive and can force out the native species. Sika deer have also caused problems in places where they have been introduced in Europe, because they have hybridized with the native red deer. In most European countries where both species occur, there are no longer any pure

Sika deer have been introduced into many countries throughout the world. Photo by Quartl.

red deer populations. In America, one fear is that sika deer could escape from game farms in western states and hybridize with elk.

Another concern is that exotic deer could bring in new diseases and parasites that are not found in native deer. On the other hand, what occurs most often is that the exotic species becomes infected with the parasites and diseases of the native population. Nonetheless, introductions of non-native deer should be discouraged. Today, almost all state and federal agencies prohibit the introduction of free-ranging exotic deer. As a result, people bringing deer into game farms or ranches, or moving stock among different states, now face more restrictive regulations than they did in the past.

Where have deer been introduced?

Exotic species of deer have been imported into numerous countries throughout the world. Australia and New Zealand have been particularly popular destinations, because neither have any native species of deer. Fallow deer, native to the Middle East, have been the most common species introduced to new locations. They have been brought into a number of U.S. states and Canadian provinces, several Caribbean islands, and various South American countries. Fallow deer now live in portions of Europe,

Russia, South Africa, Australia, New Zealand, and even Fiji. Red deer, which are native throughout much of Europe and Asia, have been introduced to the United States, several South American countries (where they negatively impact the native huemul), Morocco, South Africa, Australia, and New Zealand. Sika deer occur naturally in Japan, China, and parts of Southeast Asia. Wild sika deer populations are now established in Europe, parts of Russia, the United States (Texas, Maryland, Kentucky, and North Carolina), Australia, and New Zealand.

Two species of deer have been introduced to Hawaii, which has no native deer. Axis deer, or chital (*Axis axis*), were brought to Maui in 1959, and this species now can be found there and on the small islands of Molokai and Lanai. Black-tailed deer (*Odocoileus hemionus columbianus*), a subspecies of mule deer, were brought to Kauai two years later. Populations of both species are increasing, especially axis deer, causing damage to both native vegetation and agricultural crops.

Javan rusa deer (*Rusa timorensis*) now occur as exotics on many East Indian islands, as well as in Australia, New Zealand, several islands in the South Pacific, the Comoro Islands, and Mauritius. Sambar (*Rusa unicolor*) have been introduced to the United States (Texas, California, and Florida), Australia, and New Zealand. Native North American white-tailed deer, moose, elk, and caribou have all been introduced to other countries; whitetails now live in parts of Europe, the West Indies, Australia, and New Zealand. Even Chinese water deer (*Hydropotes inermis*) have been relocated to England, as has Reeve's muntjac. Africa is the only place where few deer have been introduced, because of the abundance of antelope and other well-established ungulates (hoofed mammals).

Fallow deer, such as these in England, are the most commonly introduced species of deer worldwide.
Photo by Hans A. Rosbach.

Why introduce deer to different places?

In the mid-1800s, one of the primary reasons to import deer was to compensate for perceived "deficiencies" in the types of native deer and to increase diversity for sport hunting. New Zealand has long been a favorite location in which to introduce deer. Like every island far from a mainland, New Zealand has no native deer species. In the 1850s, hunters and landowners brought in the first of eight introduced species: sika deer, fallow deer, red deer, North American elk, rusa deer, sambar, white-tailed deer, and moose. Of these, red deer, sika deer, sambar, and fallow deer are well established on the country's North Island, while red deer are abundant on the South Island. Most of these early introductions were done by private individuals, often landed gentry or expatriates wanting the same type of wildlife found in "the old country." Few, if any, governments permit such introductions today.

Chapter 10

Human Problems (from a deer's viewpoint)

Do people hunt and eat deer?

People have hunted and eaten deer for thousands of years. Deer have provided meat (venison), hides for shelter and clothing, and bones and antlers for tools and artwork for early Paleolithic peoples and indigenous (native) societies worldwide (see chapter 8). Deer continue to be vital to many present-day subsistence hunters (those who hunt for survival, not sport). Most people today, however, hunt deer for recreation, trophies, and as a part of enjoying the outdoors, as well as for the meat. The primary goal of most wildlife management agencies is to maintain population densities of deer within the biological carrying capacity of their habitats—that is, the ability of the land to support them. Deer populations also should be within the "cultural carrying capacity," such that people do not consider them to be too numerous. In the United States, hunters have historically been—and continue to be—the primary financial support for the management of deer through hunting licenses, fees, and federal taxes on firearms and ammunition.

Although figures vary each year, in the United States alone there are about 11 million deer hunters who are in the field an average of 12 days each season. Deer hunting annually generates approximately $25 *billion* in retail sales for equipment, travel, food and lodging, and associated activities in the United States, according to data from the International Association of Fish and Wildlife Agencies and the National Survey of Fishing, Hunting, and Wildlife-Associated Recreation, published by the U.S. Fish and Wildlife Service.

Every U.S. state has a deer season, with associated bag limits. Authorities closely monitor harvest rates to avoid harming deer populations. Deer-hunting restrictions may vary among localities within a state, depending on the population density of the deer, their age structure—the relative number of fawns, yearlings (18 months old), and adults—and overall indications of herd health. Most management agencies encourage hunters to harvest a higher percentage of females, to control the population and keep deer in balance with their habitat. States often schedule different hunting seasons, including those for hunters using rifles, shotguns, bows and arrows, crossbows, handguns, and even old-fashioned muzzleloaders. They may also offer special hunts for young or disabled hunters. Opposition to hunting is limited but vocal, and antihunter and animal rights activists often work to end or disrupt deer hunting,

White-tailed deer (*Odocoileus virginianus*) offer a snapshot of the dynamic between hunters and deer. In recent years, the whitetail population has increased dramatically throughout much of the eastern United States. One reason is that there are fewer hunters, coupled with landowners who are much less willing to let hunters onto their land, leaving naturalists and wildlife managers with the challenge of figuring out how to lower or stabilize the deer population. The average age of hunters is increasing, and states are working to maintain a multigenerational hunting community to enable adequate deer management.

Public hunting rights in European countries are generally more restrictive than in the United States. Nonetheless, since the 1960s the number of hunters in many European countries has increased, as have deer populations. As a result, annual harvest rates of red deer have risen in most countries, including Austria, Denmark, France, Germany, Hungary, Norway, Poland, Scotland, Slovenia, Sweden, and Switzerland. Deer hunting in Europe is estimated to directly generate 16 billion euros annually—about $22.9 billion—with additional environmental and social benefits. Unlike the situation in the United States, deer in Europe belong to the landowner, so landowners can regulate harvests on their properties.

Unfortunately, illegal and unregulated hunting has significantly reduced populations of many of the species of deer worldwide, now making them endangered or threatened, especially throughout Asia and Latin America. Armed conflicts in these regions have led to an increase of modern firearms into traditional hunting communities. Reducing illegal hunting to save endangered species remains a primary focus of conservationists in these regions. Trophy deer hunting in developing countries, properly regulated, could bring significant economic benefits, as well as provide critical funding for expanded conservation efforts. Adequate controls on illegal harvests, however, currently do not exist in Asia and South America. Conversely, for some species of deer that are overabundant, the loss of large predators had left public hunting as one of the few methods available to help keep populations within acceptable limits.

Are any deer species endangered?

Several species of deer are threatened or endangered, primarily in Latin America and Asia. As noted above, deer in the Southern Hemisphere often fall victim to inadequate game laws and poor law enforcement. Extensive habitat loss or fragmentation, because of agriculture and industrial development, poses another threat. Unfortunately for deer, they are fairly large and their meat is nutritious and tasty. Thus, overhunting and poaching pressure from rapidly growing human populations—especially in developing countries—have been significant factors in the decline of some deer species. In many instances, deer also face competition from livestock, predation, harm from invasive plants, and genetic deterioration from small and declining breeding populations.

Many species and subspecies of deer in Asia are endangered or threatened, with several probably on the verge of extinction. Species that are naturally restricted to small geographic areas (endemic species, or endemics) are particularly vulnerable to the threat of extinction, especially those living only on islands. In the Philippines, Visayan spotted deer (*Rusa alfredi*) and Calamian hog deer (*Axis calamianensis*) are endangered. Populations of both species are increasing, however, thanks to intensive conservation programs. Critically endangered Bawean deer (*A. kuhlii*) have benefited from conservation efforts as well, including protection from hunting, habitat management to increase the amount of browse, and a captive breeding program. The Bawean deer has the most restricted distribution of any cervid—it occurs naturally only on 77 square miles (200 km^2) of Bawean Island, near Java in Indonesia.

A subspecies of Eld's deer (*Rucervus eldii eldii*) is critically endangered because its wetland habitat in India has been lost to agriculture, as have five subspecies of sika deer (*Cervus nippon*) in China and Vietnam. Endangered Persian fallow deer (*Dama mesopotamica*), now numbering fewer than 250 individuals, are restricted to two refuges in Iran. Large-antlered muntjac (*Muntiacus vuquangensis*), limited to the Annamite Mountains of Laos, Vietnam, and perhaps eastern Cambodia, are endangered because of heavy hunting pressure throughout their range. Schomburgk's deer (*Cervus schomburgki*), which lived exclusively in swamps and grasslands in Thailand, were driven to extinction in the 1930s because of overhunting and the conversion of their habitat to agricultural lands.

In Central and South America, many of the same threats exist. Both pampas deer (*Ozotoceros bezoarticus*) and southern huemul (*Hippocamelus*

Habitat loss includes forest clearcutting (*top*), drained wetlands (*middle*), and housing developments (*bottom*).
Photos by Steve Hillebrand, USFWS (*top*), from the USFWS (*middle*), and by Claire Dobert, USFWS (*bottom*).

The endangered Bawean deer has the smallest range of any species of deer. Photo by Midori.

A Schomburgk's deer in the Berlin Zoo in 1911. The species is now extinct. Photo by Lothar Schlawe.

bisulcus) are endangered because of excessive hunting, habitat loss, and, for huemul, competition from exotic red deer (*Cervus elaphus*) and attacks by feral and domestic dogs. Pampas deer have benefited from private and state-funded initiatives to establish reserves, in addition to a captive breeding program, but they still remain at critically low densities.

Although not endangered, many other species with decreasing populations are considered vulnerable or threatened, meaning that if current population trends continue, these species will also become endangered. In South America, marsh deer (*Blastocerus dichotomus*) and northern huemul

Pampas deer (*left*) and southern huemul (*right*) are both endangered species in South America. Photos by L. Pinder, courtesy of the American Society of Mammalogists (*left*), and courtesy of JoAnne Smith (*right*).

(*Hippocamelus antisensis*) are deemed vulnerable because of hunting, logging, cattle ranching, and other factors. Five species of brocket deer are also threatened: São Paulo brocket deer (*Mazama bororo*), Mérida brocket deer (*M. bricenii*), dwarf brocket deer (*M. chunyi*), Yucatan brown brocket deer (*M. pandora*), and Ecuador red brocket deer (*M. rufina*). In addition, both the northern and southern pudu (*Pudu mephistophiles* and *P. puda*) are considered to be threatened. Thus, of the 16 species of deer in South America, 10 are either endangered or threatened.

Threatened deer species in Asia include Chinese water deer (*Hydropotes inermis*), white-lipped deer (*Przewalskium albirostris*), barasingha (*Rucervus duvaucelii*), hog deer (*Axis porcinus*), and black muntjac (*Muntiacus crinifrons*). There is currently too little information available to accurately determine the conservation status of several other species of muntjac with declining populations.

Nonetheless, there is a positive side to this story. Deer are relatively easy to breed in captivity, and they do not require as much habitat as their predators do. Deer may cause agricultural damage, but they do not cause loss of human life, as do some other large, "charismatic" species, such as tigers (*Panthera tigris*) and grizzly bears (*Ursus arctos*). Most deer adapt to people and can live within a human-modified habitat, if they have sufficient protection. And, after many decades, biologists have compiled a great deal of expertise and experience in managing deer populations.

Chinese water deer, the only species of deer without antlers, are a threatened species in Asia. Photo courtesy of Chen Min.

What is so unusual about the endangered Père David's deer?

Père David's deer (sometimes called "back-antlered deer"), which occur naturally only in China, are very unusual, because of both their physical appearance and their conservation history. Physically, this deer is a very ungainly looking animal. It has an exceptionally long tail (like a donkey); very large feet (like a cow) that are adapted to wetlands; a long, camel-like neck; and antlers that appear to be situated backward on its head. As a result, its Chinese name (*sse-pu-hsiang*) is variously translated as "none of the four," or "not a deer, not an ox, not a goat, and not a donkey," or some variation on this theme.

We are very fortunate to still be able to count this species among the living. Père David's deer originally inhabited the swamps and wetlands of northeastern China, but the species has long been extinct in the wild because of overhunting and the draining of wetlands for agriculture. The only surviving deer were those confined in the huge, walled, heavily guarded Imperial Hunting Park (Nanyuang Royal Hunting Garden) just south of Beijing. These deer were strictly off limits and out of sight to everyone outside the imperial court for hundreds of years. In 1865, a French missionary in China, Father Armand David, was allowed to look over the wall of the imperial park (after bribing the guards). Besides being a priest, Father (*Père* in French) David was an excellent naturalist. When he saw these deer, he immediately recognized them as a unique species; he speculated that they might be a new type of reindeer. A year later, with the help of the French

A Père David's deer. This somewhat ungainly looking species has a remarkable conservation history. Photo courtesy of Rory Putman.

embassy, he managed to have several samples sent back to Paris, where the renowned zoologist Henri Milne-Edwards described them scientifically as *Elaphurus davidianus*.

Once Père David's deer became more widely known to the outside world, several European zoos requested living specimens—as did the Duke of Bedford, who had imported numerous exotics from around the world to his estate at Woburn Abbey in Bedfordshire, England. Moving living specimens of Père David's deer outside of China at that time saved the species. In 1895, the Yongding River flooded and destroyed part of the wall of the imperial park. Most of the deer escaped, but they quickly perished—or were killed and eaten by starving Chinese. Five years later, the Boxer Rebellion claimed the 20 to 30 remaining Père David's deer still in China. Only the captive animals in Europe remained. Although the zoo animals failed to breed successfully, those at Woburn Abbey survived and reproduced, despite dramatic reductions in their numbers during both world wars. By the 1970s, their numbers had rebounded to over 500 deer at Woburn Abbey, with more in zoos worldwide. From these populations, authorities brought a few Père David's deer back to China in 1956, although significant reintroduction efforts did not really begin until the 1980s. Naturalists introduced 38 deer to Beijing's Milu Park in 1985, and 39 more to Dafeng Reserve, on the coast of the Yellow Sea. Additional reintroductions were made to the Tianezhou Milu Reserve in 1993 and the Yuanyang Yellow River Nature Reserve in 2002. Growth rates for all of these populations have been exceptionally high—around 20 percent annually—so even though Père David's deer remains endangered, the population is increasing.

What about other conservation efforts?

Père David's deer can be considered a conservation success story, at least thus far, because of captive breeding programs in reserves and zoos, followed by reintroduction into their native habitats. This strategy could also be used to increase the genetic diversity of additional species of deer. Other ways of helping endangered deer species recover are fairly obvious. A vital first step is to determine the number of animals and whether the population is growing or not—often a very difficult thing to do in many areas. Decreasing or eliminating populations of predators, such as feral dogs or invasive exotic species, may be a part of an overall recovery plan as well.

Because overhunting of many species remains a serious problem, the enforcement of existing regulations to reduce hunting and poaching is essential, as is strengthening existing nature reserves and establishing new ones. In practice, however, such initiatives can be difficult, given existing socioeconomic and political conditions. In all such conservation efforts, public education, understanding, and support is essential for the long-term success of any program to reestablish and protect deer and their habitats.

Why are some species endangered while others reach pest proportions?

Some species have naturally small ranges, or they face greater competition from other hoofed mammals (ungulates) within their range. Sometimes the introduction of exotic species has reduced the abundance or distribution of native deer. Just as often, a lack of strong management, unregulated hunting, and the destruction of critical habitats can lead to conservation crises. This was certainly the case for white-tailed deer in the United States in the late 1800s and early 1900s, when they became extirpated (locally extinct) in many regions. In Illinois, for example, whitetail populations in the 1850s were so low that the state imposed hunting restrictions for the first time. Periodic limits remained in place until 1901, when a complete ban on deer hunting was initiated. There was no deer hunting in Illinois again until 1959. During this 58-year interval, authorities translocated white-tailed deer—brought them from the upper Midwest and New England to Illinois—to increase their population size. Similar efforts to aid the recovery of white-tailed deer also occurred in other states. Today, whitetails are overabundant in many areas, meaning that their populations have grown beyond the carrying capacity of their habitats.

Recovery efforts are ongoing for many deer species around the world that are currently threatened or endangered. But as human populations continue to increase, especially in Latin America and Asia, deer populations and their habitats continue to decline—especially where there is failure to regulate or manage herds.

How will deer be affected by global warming?

Global warming may impact all living organisms, and deer are no exception. Increased warming and higher levels of carbon dioxide will change patterns of temperature and precipitation, thus pushing frost lines and snow to higher elevations—and farther toward the poles. Climate change may affect deer and their typical environments by several means: changing the composition, quality, and quantity of vegetation; increasing the competition between deer species because of habitat changes; shifting predation rates through changes in the snowpack; increasing the spread of parasites and diseases that attack deer; altering the optimal times for giving birth; and (perhaps) increasing the occurrence and severity of droughts and wildfires, among other factors. Global warming will harm the growth, reproduction, population abundance, and distribution of some species, but other species may benefit from such changes. Among deer species overall, generalist herbivores are likely to be less affected than specialist browsers or grazers.

It is difficult, however, to predict how each species will respond to the array of interacting climate variables. Their responses probably will vary locally and regionally throughout the many geographic ranges of deer. Generally, species that are the most mobile and can adapt to new foods and habitat conditions may benefit from global warming. On the other hand, the migration of species such as caribou (*Rangifer tarandus*), North American elk (*Cervus elaphus*), or mule deer (*Odocoileus hemionus*) could be hampered in some areas by human developments.

Rising temperatures will alter the structure, composition, and nutritional value of the plants in the forest understory—that is, the shrub and grass layer deer depend on most for food and shelter. If suitable understory vegetation is reduced, or if the places in which it occurs becomes more fragmented, the nutritional stress on deer will increase, especially in ecosystems that are already hot and dry. Conversely, in some regions the understory vegetation may increase. Changes to northern coniferous (evergreen) forests, alpine and meadow habitats, and deciduous forests (those with trees that lose their leaves in the fall) also will affect deer populations, but alterations will be positive in regions where warming decreases the length and severity of winter conditions.

Changes in local habitats and in the broader ecosystem could influence competition for food among various species of deer, thus affecting the reproduction and abundance of some species. Rates of predation may also change. In regions with wolves (*Canis lupus*), predation rates on deer usually are linked with the severity of winter temperatures and snow depth. Global warming could reduce deer mortality (deaths), and thus increase their population densities, in areas where milder winters translate into less hunting success for wolves. If warmer winter temperatures increase snowfall, however, as some climate models predict, deer populations could decline.

Among the numerous kinds of parasites and diseases in deer, white-tailed deer carry a roundworm (nematode) called meningeal worm (*Parelaphostrongylus tenuis*). A closely related and biologically similar group of roundworms (genus *Elaphostrongylus*) occurs in deer in Europe and Asia. Although *Parelaphostrongylus* does not significantly affect white-tailed deer, it does causes "moose sickness," which can be fatal in moose (*Alces alces*), elk, and other deer. This parasite needs terrestrial gastropods (snails and slugs) as secondary hosts to complete its life cycle. As global temperatures rise, changes in the density and distribution of white-tailed deer and snails, and a longer time period in which the parasite remains infectious, could all interact and lead to increased mortality for other deer species.

This scenario certainly applies to the many other diseases and parasites that infect deer. One example is the winter tick, *Dermacentor albipictus*. This tick infests various species of deer, but moose are most severely affected, with up to 30,000 or more ticks attaching themselves to one animal. These ticks can cause moose to lose weight, and up to 50 percent of their hair, during the winter, leading to extensive heat loss. In some areas, hundreds of moose probably die because of these overwhelming tick infestations. Increasing temperatures may well increase the abundance of ticks.

Several deer species may already be feeling the effects of global warming. Moose, for example, become heat stressed if the ambient (outside) air temperatures become too high; 57° Fahrenheit (14° Celsius) in the summer is the upper limit before their body condition and reproductive rates decline. In northern Minnesota, moose populations have decreased by 90 percent in the past 20 years, a trend attributed in part to rising temperatures. Conversely, red deer in Siberia have extended their range northward, possibly because of higher ambient temperatures and reduced snowpack. In many European countries, populations of red deer, roe deer (*Capreolus capreolus*), moose, and other species of deer have increased in density and distribution in the past 40 to 50 years, probably in part because of climate change. Likewise, climate models predict that warmer summer temperatures will cause significant increases in the populations of North American elk.

Chapter 11

Deer in Art and Literature

What roles do deer play in art, religion, mythology, and popular culture?

Large mammals—including deer—have been prominently portrayed since the beginnings of human history. In addition to literal representations, deer have played a significant figurative role, as part of the symbolism and metaphors found in art, mythology, and culture.

Paleolithic people 20,000 to 35,000 years ago produced art as portable images painted or carved on small stones, bones, ivory, and antlers. They also made the more familiar non-portable, detailed, and sophisticated paintings or carvings on the walls of caves.

Deer were objects of art and mythology for the earliest Hittite (people from Anatolia, or Asia Minor, in what is now western Turkey) and Egyptian cultures, the Greeks, the Celts, the Hindus, and others throughout the ancient world. Stags (males) were usually believed to have supernatural size and powers, while hinds (females) often were credited with befriending or nursing saints, kings, or other heroes who were in need.

Deer were common mythical symbols in early Asian culture as well, and they were believed to remove the spirits of the dead. For example, in Japanese Shinto religious beliefs, sika deer (*Cervus nippon*) are messengers to the gods. In China, deer symbolize health and long life (longevity). A spotted deer is believed to accompany the god of longevity, and deer in general are credited with having strong medicinal value.

For the Huichal Indians of Mexico, deer represented an important sacrificial animal. In medieval Europe, deer had a predominant role in literature and art, especially in hunting scenes, and they often graced coats-

Top, **A prehistoric painting of a deer, dating from 12,000 years ago, in the La Pasiega cave, located in Puente Viesgo, Spain.** ***Top right***, **A drawing of a giant elk,** ***Megaloceros***, **from a cave painting in Lascaux, France, dating back 20,000 years ago.** ***Bottom***, **Cave art of a deer, from approximately 6,000 years ago, in the Parque Natural de la Sierra y los Cañones de Guara in Huesca, Spain.** Sketch by José-Manuel Benito Álvarez (*top*), photos by HTO (top right), and Hugo Soria (*bottom*).

of-arms. Deer continue to be an important part of popular culture in art and literature throughout the modern world, and they are well known to children through characters such as Bambi and Rudolph the Red-Nosed Reindeer.

How have deer been incorporated into literature and poetry?

Deer have been a major part of literature and poetry since the dawn of the written word. Deer and other animals have served as fictional fig-

Pottery decorated with images of deer, dating from about 550 B.C., from Lydos, Greece. Photo by Jastrow.

A painting by an unknown court artist from the *Album of the Yongzheng Emperor in Costumes.* From the Royal Academy of Arts, London.

ures and symbols in fables (brief stories in which animals exhibit human qualities) and allegories (works that have a symbolic, rather than a literal meaning)—usually to bring a definite moral lesson home to readers, so we can better understand ourselves. Deer have also been the focus of early guides on hunting, anatomy, and medicine, as well as in works on natural history.

A painting entitled *The Stag and the Horse*, by Wenzel Hollar (1607–1677).

From the University of Toronto.

Sumerian writings (from southern Mesopotamia, in what is now Iraq) on clay tablets, dating from 5,000 years ago, are among the world's oldest surviving compositions. Although these earliest writings were utilitarian, tracking such practical matters as economic or administrative records, non-utilitarian texts appeared somewhat later. Deer (stags) became symbols in many of these latter Sumerian works. The Sumerian god Enki, for example, travels the river "most delightfully" on his barge, the *Stag of Abzu*. This stag would have been a fallow deer (*Dama dama*). Likewise, the goddess Inana "stands among stags in the mountain tops, she possesses fully the divine powers." The Rig Veda, written in the ancient Sanskrit language, is the oldest sacred book in the Hindu religion. Dating from 1700 B.C., it is made up of over 1000 hymns, some of which make reference to deer, including Hymn 39: "To the Maruts" (storm gods): "You have harnessed the spotted deer to your chariots, a red one draws as leader; even the earth listened at your approach, and men were frightened."

Deer also are mentioned several places in the Bible, including 2 Samuel 2:18, 1 Chronicles 12:8, and 1 Kings 4:23. In Deuteronomy 14:5, the dietary laws list animals people are permitted to eat, including the "hart, and the roebuck, and the fallow deer." In chapters from the Song of Songs, deer are considered to be symbols of beauty and grace:

Deer in heraldry. Countless European coats-of-arms feature deer, such as the three examples shown here. *Left*, Castillo Nuevo, Spain. *Middle*, Åland, Finland. *Right*, Dozwil, Canton of Thurgau, Switzerland. Photos by Miguillen (*left*), Wikimedia Commons (*middle*), and D. Weidmann (*right*).

> Chapter 2:9 My beloved is like a roe or a young hart: behold, he standeth behind our wall, he looketh forth at the windows, shewing himself through the lattice.
>
> Chapter 8:14 Make haste, my beloved, and be thou like to a roe or to a young hart upon the mountains of spices.

Early fables abound with tales of deer. The best-known ones are credited to the legendary author Aesop, who lived in Greece in the sixth century B.C. These fables—featuring animals, all with human characteristics—include clear moral lessons. For example, in "The Stag at the Pool," a deer is admiring his antlers while bemoaning his spindly legs. When a hunter (or lion, or pack of dogs, depending on the re-telling) approaches, the deer flees easily, but his antlers entangle him in the forest vegetation, and he perishes when overtaken by his pursuers. The moral: do not value beauty at the expense of what may be most useful to you. Likewise, in "The Sick Stag," it is easy to understand the message that uncaring friends can be more harmful than helpful:

> A sick stag lay in a corner of his pasture. His friends came to inquire about his health and each who came ate some of the food put there for the stag. He eventually died, not from illness, but from lack of food.

Other allegorical fables attributed to Aesop that involve deer are "The Stag in the Ox-Stall," "The One-Eyed Doe," "The Stag and a Lion," and "The Fawn and His Mother." Each has since served as the basis for numerous, similar fables by others.

A group of tales, originally written in the second century A.D., were collected and circulated in the Middle Ages under the title *Physiologus*. They were presented in illustrated volumes, called bestiaries, and numerous translations in Latin, French, English, German, and other languages appeared throughout the medieval period. The factual aspects of these stories—descriptions and other natural history information—were, however, often secondary to the religious principles (doctrines) they contained. In other words, the stories were allegories, giving each animal, plant, or inanimate object (such as a rock) a certain human characteristic—majesty, cunning, or productivity, for example—that served as a symbol or a metaphor to deliver a moral lesson.

These allegorical tales often highly praised the powers of deer. Consider, for instance, the following from Isidore of Seville, a seventh-century Spanish bishop and noted scholar:

> Deer are the enemy of snakes. When deer are ill or weak they draw snakes out of their holes with the breath of their nostrils and eat them, overcoming their poison and thus renewing themselves. They also use the herb dittany as a medicine; eating it causes arrows that have struck them to fall out. When their ears are erect their hearing is sharp, but when their ears are down they cannot hear at all.

Snakes represented the devil, and stags—a symbol for Christ—destroyed evil. Likewise, Isidore described how deer crossing a wide river swim in a line, with each animal resting its head on the rump of the one in front. The moral principle? Good Christians crossing over to a spiritual life should assist others who may be weaker and more tired.

Throughout the twelfth to the seventeenth centuries, numerous works depict deer artistically, but many also portray them from more utilitarian, descriptive standpoints, such as hunting guides or medicinal handbooks. In his extensive volume entitled *Man and the Natural World*, Keith Thomas clearly shows that in England and much of Europe during this period, "the long-established view was that the world had been created for man's sake and that other species were meant to be subordinate to his wishes and needs." Strong theological (religious and philosophical) arguments were given to support the idea that every animal was "intended to serve some human purpose, if not practical, then moral or aesthetic." A clear example that includes deer is from Dame Juliana Berners, in *The booke of hauking, huntyng, and fysshyng, with all the properties and medecynes that are necessary to be kept*—written in the early 1400s—where she discusses "beastes of the chase," including "the bucke, the doe . . . the raynder, and the elke." A search of the website Early English Books Online using the keyword

A manuscript page from a Latin bestiary, *Omnibus animantibus*, by Isidore of Seville, MS 254, housed in the Fitzwilliam Museum, Cambridge University. Note deer swimming with their chin on the animal in front. Used with permission of Cambridge University.

"deer," for example, yields 4,019 references to deer in 1,257 early English publications, beginning in 1473, while the keyword "stag" results in 1,868 references in 785 works. Numerous volumes suggest medicinal remedies. For example, in a 1663 medical treatise, Robert Bayfield recommends "a piece of the hoof of an Elk, which indeed is famous for its specifical vertue against the Falling-sickness." Other early literature discusses basic natural history. In 1666, Englishman George Alsop, in his work on a newly settled American colony, entitled *A Character of the Province of Mary-land . . .*, alluded to white-tailed deer (*Odocoileus virginianus*):

> The Deer here neither in shape nor action differ from our Deer in England. . . . They are also mighty numerous in the Woods, and are little or not at all affrighted at the face of a man, but . . . though their hydes are not altogether so gaudy to extract an admiration from the beholder, yet they will stand (almost) till they be scratcht.

During the Renaissance and Neoclassical periods, many of these seventeenth- and eighteenth-century writers and artists frequently depicted deer, only now the dominant theme became a much more sympathetic portrayal of these animals as sensitive, vulnerable, feeling creatures. As noted by Keith Thomas, by this time the "confident anthropocentrism" of Tudor England—the idea that animals were simply useful for people—had "given way to an altogether more confused state of mind. The world could no longer be regarded as having been made for man alone, and the rigid barriers between humanity and other forms of life had been much weakened." An example is the *Poem Scroll with Deer*, an early seventeenth-century masterpiece from Renaissance Japan. A collaboration between the famed calligrapher Hon'ami Koetsu and painter Tawaraya Sotatsu, the *Deer Scroll* (as it is commonly called) illustrates an anthology, or collection, of 28 Japanese poems about autumn with simplified, elegant, and idealistic images of sika deer in a variety of poses, drawn using silver and gold pigments.

Although not mentioned as often as birds, deer do appear among the more than 4,000 references to animals in the works of Shakespeare. According to Open Source Shakespeare online, the word "deer" occurs 43 times in 40 lines among 20 different works. Deer are mentioned most often, and by various characters, in *Love's Labour's Lost* and *As You Like It*. In Act III of *Julius Caesar*, Antony draws upon the image of a hunted, fallen deer (hart) to symbolize the death of Caesar:

> Pardon me Julius! Here was thou bay'd, brave hart,
> Here didst thou fall; and here thy hunters stand,
> Sign'd in thy spoil, and crimson'd in thy lethe.
> O world, thou wast the forest to this hart;
> And this, indeed, O world, the heart of thee.
> How like a deer, stricken by many princes,
> Dost thou here lie!

The allegorical theme of fallen nobles or other tormented heroes as stalked, suffering, and eventually slaughtered deer is well represented in the works of many other fifteenth- to nineteenth-century writers. Anne Elizabeth Carson, in her essay "The hunted stag and the beheaded king," explored how poets of the time treated the execution of King Charles I of England in 1649. She noted: "Just as a helpless beast inevitably succumbs to pain and torment, so it is with a human being: a woman, a man, even a king. . . . The figure of the stag or fawn becomes the metaphoric incarnation of both the poet and the nation's despair, and such a creature invariably comes to embody the deceased king." The symbol of the hunted stag, representing King Charles, appears in such works as Margaret Cavendish's "Hunting of a Stag" in 1664 and John Denham's "Cooper's Hill" in 1668. The theme of King Charles I as a hunted deer still resonated 200 years later. Consider these lines from "The Execution of Montrose," written by the Scottish poet William Aytoun in 1863, commemorating the hanging of James Graham, Marquis of Montrose, who was a supporter of the king:

The grim Geneva ministers
With anxious scowl drew near,
As you have seen the ravens flock
Around a dying deer.

Similarly, Percy Bysshe Shelley, in his poem "Orpheus" (published in 1862), compares the title character to a pursued deer:

Awhile he paused. As a poor hunted stag
A moment shudders on the fearful brink
Of a swift stream—the cruel hounds press on
With deafening yell, the arrows glance and wound,—
He plunges in

William Wordsworth's 1849 poem, "The Russian Fugitive," also contains clear allusions to the tormented heroine as a hunted deer:

A shout thrice sent from one who chased
At speed a wounded deer,
Bounding through branches interlaced,
And where the wood was clear.
The fainting creature took the marsh,
And toward the Island fled,
While plovers screamed with tumult harsh
Above his antlered head;
This, Ina saw; and, pale with fear,
Shrunk to her citadel;
The desperate deer rushed on, and near
The tangled covert fell.
Across the marsh, the game in view,
The Hunter followed fast,
Nor paused, till o'er the stag he blew
A death-proclaiming blast;
Then, resting on her upright mind,
Came forth the Maid—"In me
Behold," she said, "a stricken Hind
Pursued by destiny!

Although many writers throughout history have been sympathetic to animals, communion (or fellowship) rather than conflict with animals is a dominant theme throughout much of the Romantic period. As noted by David Perkins in his essay "Wordsworth and the polemic against hunting," in England animals were "thought to be capable of happiness and suffering . . . viewed as individuals, each with [its] own unique personality and life history. Moreover, animals had been endowed by nature with rights." Thus many poets of the time attacked hunting and "field sports" as barbaric and cruel, with animals—often deer—gaining our sympathy and the hunter our scorn and indignation. This can be seen in the works of poets such as Keats, Byron, Thomson, Cowper, and Coleridge, where antihunting sentiments abound. In his masterpiece "The Seasons," published in 1730, Scottish poet James Thomson gives us this picture of the final, agonizing moments of a hunted deer:

> The big round tears run down his dappled face,
> He groans in anguish; while the growing pack,
> Blood-happy, hang at his fair jutting chest,
> And mark his beauteous chequered Sides with gore.

Likewise, from Wordsworth's tragic play *The Borderers*:

> never choose to die,
> But some one must be near to count his groans.
> The wounded deer retires to solitude,
> And dies in solitude: all things but man,
> All die in solitude.

Deer continue to be portrayed in art and literature today in a variety of forms, much as they have been throughout history. There are countless contemporary books and articles on deer hunting and natural history; deer occur prominently in much of children's literature, including young-adult classics like Marjorie Kinnan Rowling's *The Yearling*; and images of deer grace everything from serious art to company logos, such as those for John Deere and Caribou Coffee.

Much of contemporary writing continues the often-contradictory views of deer and other animals that are seen throughout history: humans brutalizing animals, juxtaposed with our awe at and communion with them. Sadly, however, given much of modern life, contemporary poets such as James Tate illustrate our failed, ambivalent, and hypocritical relationship with nature, as in his "The Pet Deer," published in 1970. The princess (representing people) causes distrust in the deer (representing nature); our lack of understanding is illustrated in the last line:

> The Indian Princess in her apricot tea gown
> moves through the courtyard teasing the pet deer
> as if it were her lover. The deer, so small and
> confused, slides on the marble as it rises on its hind legs
> towards her, slowly, and with a sad, new understanding.
> She does not know what the deer dreams or desires.

The grace and beauty of deer are captured in these modern wood sculptures. Photo by George Feldhamer.

One final example of this theme of failed communion and stewardship with nature and animals—with a dead, pregnant doe symbolizing the difficult and conflicting decisions we must make in our relationships with other animals—is seen in the poignant poem by William Stafford, entitled "Traveling through the Dark":

Traveling through the dark I found a deer
dead on the edge of the Wilson River road.
It is usually best to roll them into the canyon:
that road is narrow; to swerve might make more dead.
By glow of the tail-light I stumbled back of the car
and stood by the heap, a doe, a recent killing;
she had stiffened already, almost cold.
I dragged her off; she was large in the belly.
My fingers touching her side brought me the reason—
her side was warm; her fawn lay there waiting,
alive, still, never to be born.
Beside that mountain road I hesitated.
The car aimed ahead its lowered parking lights;
under the hood purred the steady engine.
I stood in the glare of the warm exhaust turning red;
around our group I could hear the wilderness listen.
I thought hard for us all—my only swerving—,
then pushed her over the edge into the river.

Chapter 12

"Deerology"

Who studies deer?

There are several kinds of professionals who study deer. In North America, most of the people responsible for deer management work within state, provincial, and federal governments. Some experts, however, work as private consultants or as managers for large landowners. Deer managers are responsible for estimating deer numbers; planning and carrying out programs for increasing, decreasing, or stabilizing deer populations; and responding to complaints and issues around deer-human conflicts. As a part of this latter group, conservation police officers (game wardens) are charged with enforcing regulations regarding the legal harvest of deer. Deer managers usually have an advanced degree (an M.S. or a Ph.D.) in wildlife ecology or management, and often they are certified as wildlife biologists by The Wildlife Society, the professional organization to which many of them belong. Their training includes courses in nutrition; habitat use; the estimation of population size, growth and reproduction; wildlife law; habitat manipulation; and genetics. The Wildlife Society includes not only deer managers, but scientists interested in most game and non-game species and their habitats, as well as professionals involved in other aspects of management, such as policy formulation or the human dimensions associated with deer.

In Europe, deer managers are employed by private individuals, or by forestry or hunt clubs that own the land where the deer reside. In these countries deer belong to the landowner, and the animals are regulated according to his/her goals. In the United States, deer are not the private property of landowners, but landowners may hire experts to manage habitat and

State and federal conservation officers help enforce wildlife laws. Photo by John and Karen Hollingsworth, USFWS.

to regulate hunts on their land. In the developing world, there are few organizations specifically devoted to deer. Depending on the recent history of the country (such as whether it was a colony of a developed country), deer managers may be employed within government agencies or by private landowners.

Deer are also studied by scientists who are employed within universities or in state, provincial, and federal agencies. These scientists may be interested in providing practical information for deer managers, or in studying deer as one important component within an ecosystem or a line of evolutionary descent (phylogeny). Research scientists might study ecology, morphology (physical characteristics), genetics, physiology (how living organisms function), behavior, or other disciplines, focusing on deer as their study species. Teachers within colleges and universities will use deer to illustrate principles of ecology, management, or conservation to their students.

Some scientists and managers are interested in deer because of what they eat and who eats them. Many large carnivores (meat-eaters) are dependent on deer as a basic prey; thus deer numbers and their distribution are critical for managers of carnivore populations. For example, the conservation of tigers (*Panthera tigris*) depends on prey availability within their habitats, and the tigers' most common prey are deer, wild bovids, and pigs.

Foresters need to replace forests being harvested for wood products, or to plant trees as part of restoration efforts, and the probability of seedlings growing into saplings (young trees) is dependent on deer numbers and distributions. If the population densities of deer are too high, they could destroy regenerating forests (see chapter 9).

Deer are also of interest to conservationists. Most species of deer fall into the two extremes of conservation: too few or too many. For several species of deer, too great a population density can have a negative impact on the habitat they occupy. This is the case for white-tailed deer (*Odocoileus virginianus*), red deer (*Cervus elaphus*), fallow deer (*Dama dama*), and sika deer (*Cervus nippon*) in many of the places where they live. Conversely, more than 30 species are considered to be threatened or endangered by human activities (see chapter 10). These species need protection, as well as management strategies similar to those for more well-known conservation symbols, such as tigers and giant pandas (*Ailuropoda melanoleuca*). The conservation of deer species is the responsibility of either governmental agencies or non-governmental organizations, such as the Wildlife Conservation Society, the World Wildlife Fund, or the Nature Conservancy. An umbrella organization for the conservation of deer is the International Union for Conservation of Nature (IUCN) Deer Specialist Group, which brings together deer experts concerned with all issues related to deer conservation.

With the possible exception of birds, deer are the group of animals studied most by the general public. Due to a long history of hunting and cultural practices, there is a wealth of indigenous (native or ingrained) knowledge in both traditional and modern societies. Deer experts can be found

Young stag (male) fallow deer sparring. Population densities of fallow deer are often high in many parts of their range. Photo courtesy of Jackie Pringle.

in almost every hunting camp and rural village. They may be difficult to identify, due to imitators who pretend to know as much, but the knowledge possessed by some local practitioners rivals anything collected by the "experts."

Which species are best known?

The species of deer most well known within western society are white-tailed deer and elk (*Cervus elaphus*) in North America, and European roe deer (*Capreolus capreolus*) and red deer in Eurasia. Within specific cultures, other deer species are better known. Eld's deer (*Rucervus eldii*) are a part of cultural customs throughout Southeast Asia. Caribou (*Rangifer tarandus*) are essential for the cultural practices of Laplanders in northern Scandinavia, and they are central to many Inuit societies in Arctic regions. Sika deer are part of religious practices in Japan; they have been domesticated for centuries, primarily for their antler velvet. Père David's deer (*Elaphurus davidianus*), the official deer of the emperor of China, have a very complex conservation history (see chapter 10). Within the local cultures where they are abundant, many of these deer species represent fertility, wealth, and the natural world.

Most North Americans are familiar with white-tailed deer. They are the central focus of hunting, the main cause of deer/vehicle collisions, the primary host of certain diseases that are transmittable to humans, and the principal reason for people's complaints about agricultural and forest damage. White-tailed deer have the highest current population numbers of any

Most people easily recognize North American elk. Photo by Gary Zahm, USFWS.

deer species (about 30 million) and the largest geographic range of any cervid—extending from Canada south to northern South America. Mule deer (*Odocoileus hemionus*) are more common in parts of western North America, but white-tailed deer dominate both scientific and general reports about deer. Moose (*Alces alces*) and caribou are circumpolar (found throughout the region within the Arctic Circle), but they are not as widely distributed as whitetails.

In North America, elk and white-tailed deer are the symbol of several well-known companies, such as Hartford Financial Services and John Deere. Red deer, roe deer, and moose are symbols on the family crests of many of the landed gentry in Europe. Red deer are dispersed so widely and are so variable that many people don't realize that North American elk and European red deer are the same species; over 25 subspecies of red deer are recognized across Eurasia.

Deer are important elements in art and literature throughout the world (see chapter 11), and popular culture has given us some famous individual deer. The most well-known is Bambi, from the Walt Disney movie by the same name. In the film, Bambi seems to be a white-tailed deer, but the movie is based on a European book, where Bambi is a roe deer. A second notable deer is Bullwinkle, a talking moose in an international cartoon series called "Rocky and Bullwinkle." A third, and probably the most well-recognized deer in North America, is Rudolph the Red-Nosed Reindeer, the focus of a famous children's movie and Christmas song.

Which species are least known?

Most people are not aware that a total of about 50 species of deer inhabit the world. Least known are the smaller species, such as southern pudu (*Pudu puda*), muntjac (genus *Muntiacus*), tufted deer (*Elaphodus cephalophus*), and brocket deer (genus *Mazama*). Few people have heard of the small island deer of the Indo-Pacific, such as Visayan spotted deer (*Rusa alfredi*), Philippine deer (*Rusa marianna*), or Javan rusa deer (*Rusa timorensis*). There are several species of brocket deer and muntjac where researchers do not know enough about them to even guess at their status in the wild. Examples in Asia include Fea's muntjac (*Muntiacus feae*) and Gongshan muntjac (*Muntiacus gongshanensis*). Some experts recognize Sumatran muntjac (*Muntiacus montana*) as a distinct species, while others consider them to be a variant of red muntjac (*M. muntjak*). Regardless, there are too few known specimens of red muntjac to determine their population or taxonomic (scientific classification) status. In South America, not much data exist on several species of brocket deer: red brocket deer (*Mazama americana*),

Mule deer occur throughout much of western North America. Photo by Tupper Ansel Blake, USFWS.

southern brocket deer (*Mazama nana*), and Central American red brocket deer (*Mazama temama*).

How do scientists tell deer apart?

Scientists use both morphology (physical characteristics) and genetics to distinguish deer from other species of mammals, and from each other (see chapter 1). Taxonomically, deer (family Cervidae), as well as some other ungulates (hoofed mammals), share a cloven hoof—where only the third and fourth digits make contact with the ground—and a lack of upper incisors. Deer, and other ruminants (plant-eating mammals that chew and then re-chew their food), have a characteristic, four-chambered stomach (see chapter 7). The presence of antlers, however, make deer different from other ungulate species. Antlers are produced annually, and are composed of bone with a layer of velvet, a tissue with many small blood vessels that covers the antlers as they grow and supplies them with blood (see chapter 2). Each year antlers, which develop prior to the mating season (rut), become hard by the time the mating season arrives, which allows sparring between males. Most species shed their antlers once the rut is over, but caribou retain their antlers outside of the mating season, and both sexes have them, which is unusual. Chinese water deer (*Hydropotes inermis*) are an exception among deer species, as they lack antlers. Antler configuration, coloration, size, and bone structure all vary among the different species of deer. Foren-

A female (hind) chital, also called an axis deer. Photo courtesy of Jackie Pringle.

sic scientists also use hair structure and bones to distinguish species of deer. Deer species differ in the arrangement of their scent glands, although all have facial glands. All deer lack gallbladders.

Nowadays, researchers can tell deer species apart genetically, using a sample of blood, tissue, or hair. Even the exterior surface of deer pellets (fecal droppings) can contain enough DNA from the digestive tract for genetic analysis. Most studies that differentiate deer species currently use DNA—generally the cytochrome *b* gene (a section of mitochondrial DNA, which is found in one portion of a cell), although nuclear DNA sequences (the ways in which elements are arranged in a DNA molecule in another part of a cell, the nucleus) have also been used. Species are arranged on a phylogenetic tree (a branched diagram showing the evolutionary relationships among species), based on the number of substitutions in the nucleotides (the structural units in DNA) between the sequences being compared. Because the rate of nucleotide substitution is known for this segment of the DNA, the time periods when the various species evolved and diverged from each other can also be estimated.

It is fascinating that the phylogenies (relationships) constructed on the basis of physical characteristics (morphology) are not always the same as those constructed from genetic data. Populations that appear to be similar morphologically have enough genetic differences to warrant being considered separate species. This has happened for species within the genus *Mazama* in South America; much of the "discovery" of these cryptic (apparently identical, but genetically distinct) species has come about through genetic analyses. For species where populations differ in coat color and antler configurations, such as red deer, genetic analyses have revealed that the 25 morphological subspecies form 4 groups (clades) whose genetic sequences are highly similar. Lastly, extinct and living deer species can be compared and placed within the same phylogenetic tree. For example, the recently extinct Schomburgk's deer (*Rucervus schomburgki*) appears to be closely related to both chital (*Axis axis*) and barasingha (*Rucervus duvaucelii*), even though these two species differ in appearance, both to each other and to Schomburgk's deer.

Appendix A

Deer of the World

Order Artiodactyla

Family Cervidae

Scientific Name	Common Name	General Location
Subfamily Capreolinae (or Odocoileinae)		
Alces alces	Moose	North America, Siberia, N. China, Europe
Blastocerus dichotomus	Marsh deer	N. South America
Capreolus capreolus	European roe deer	Europe, Russia, Middle East
Capreolus pygargus	Siberian (eastern) roe deer	Russia, China, Mongolia, Korea
Hippocamelus antisensis	Northern huemul (taruca)	N. South America
Hippocamelus bisulcus	Southern huemul (guemal)	S. Chile, S. Argentina
Mazama americana	Red brocket deer	N. South America
Mazama bororo	São Paulo brocket deer	Brazil
Mazama bricenii	Mérida brocket deer	W. Venezuela
Mazama chunyi	Dwarf brocket deer	Bolivia, Peru
Mazama gouazoubira	Brown brocket deer	N. South America
Mazama nana	Southern brocket deer	Argentina, Brazil, Paraguay
Mazama pandora	Yucatan brown brocket deer	C. Mexico
Mazama rufina	Ecuador red brocket deer	Ecuador, Colombia
Mazama temama	Central American red brocket deer	Central America
Odocoileus hemionus	Mule deer	W. North America
Odocoileus virginianus	White-tailed deer	North, Central, and South America
Ozotoceros bezoarticus	Pampas deer	South America
Pudu mephistophiles	Northern pudu	N. South America
Pudu puda	Southern pudu	S. South America
Rangifer tarandus	Caribou (reindeer)	North America, Europe, Russia
Subfamily Cervinae		
Axis axis	Chital (axis deer)	India, Nepal, Sri Lanka
Axis calamianensis	Calamian deer	Philippines
Axis kuhlii	Bawean deer	Indonesia

Axis porcinus	Hog deer	Asia
Cervus elaphus	Red deer	Eurasia, North Africa
	North American elk (wapiti)	W. North America
Cervus nippon	Sika deer	Japan, China, E. Russia
Dama dama	Fallow deer	Middle East, widely introduced
Dama mesopotamica	Persian fallow deer	Iran, Israel; formerly in the Middle East
Elaphodus cephalophus	Tufted deer	S. China, Tibet, N. Myanmar
Elaphurus davidianus	Père David's deer	originally from China
Muntiacus atherodes	Bornean yellow muntjac	Borneo
Muntiacus crinifrons	Black muntjac	E. China
Muntiacus feae	Fea's muntjac	Myanmar, Thailand, S. China, Laos
Muntiacus gongshanensis	Gongshan muntjac	S. China, Tibet, Myanmar
Muntiacus muntjak	Red muntjac	Asia, Indonesia
Muntiacus puhoatensis	Puhoat muntjac	Vietnam
Muntiacus putaoensis	Leaf deer	N. Myanmar
Muntiacus reevesi	Reeve's muntjac	S.E. China, Tibet
Muntiacus rooseveltorum	Roosevelt's muntjac	Laos
Muntiacus truongsonensis	Annamite muntjac	Laos, Vietnam
Muntiacus vuquangensis	Large-antlered muntjac	Laos, Vietnam
Przewalskium albirostris	White-lipped deer (Thorold's deer)	China
Rucervus duvaucelii	Barasingha	India, Nepal
Rucervus eldii	Eld's deer	Asia
Rusa alfredi	Visayan spotted deer	Philippines
Rusa marianna	Philippine deer	Philippines
Rusa timorensis	Javan rusa deer	Indonesia
Rusa unicolor	Sambar	India, Sri Lanka, Asia
Subfamily Hydropotinae		
Hydropotes inermis	Chinese water deer	China, Korea

Appendix B

Deer Conservation Organizations

The following are examples of organizations that deal with deer conservation management.

American Museum of Natural History
    http://www.amnh.org
American Society of Mammalogists
    http://www.mammalsociety.org
Arizona Deer Association
    http://www.azdeer.org
Association of Zoos and Aquariums
    http://www.aza.org
Berryman Institute
    http://www.berrymaninstitute.org
Boone and Crockett Club
    http://www.boone-crockett.org
California Deer Association
    http://www.caldeer.org/index.html
CITES (Convention on International Trade in Endangered Species of Wild Fauna and Flora)
    http://www.cites.org
Conservation International
    http://www.conservation.org/discover/Pages/about_us.aspx
CTE (Center for Transportation and the Environment)
    http://itre.ncsu.edu/cte
Dallas Safari Club
    http://www.biggame.org
The Deer Initiative
    http://www.thedeerinitiative.co.uk
Deer-Vehicle Crash Information Clearinghouse
    http://www.deercrash.com
Defenders of Wildlife
    http://www.defenders.org
European Wildlife
    http://www.eurowildlife.org
Expertise Centre for Biodiversity and Sustainable Development
    http://www.ecnc.org

The Field Museum
http://www.fieldmuseum.org
IUCN (International Union for Conservation of Nature)
http://www.iucn.org
IUCN Deer Specialist Group
http://www.iibce.edu.uy/DEER/en1.htm
Key Deer Protection Alliance
http://www.keydeer.org/default.htm
The Land Conservancy of BC
http://blog.conservancy.bc.ca
Mule Deer Foundation
http://www.muledeer.org/about/index.html
National Trust (England)
http://www.nationaltrust.org.uk/main
National Wildlife Federation
http://www.nwf.org/Wildlife.aspx
Natural England
http://www.naturalengland.org.uk
Natural Resources Defense Council
http://www.nrdc.org/about
The Nature Conservancy
http://www.nature.org
North American Deer Farmers Association
http://www.nadefa.org
Northern Ireland Environment Agency
http://www.doeni.gov.uk/niea
Operation Wallacea
http://www.opwall.com
Pope and Young Club
http://www.pope-young.org/default.asp
Quality Deer Management Association
http://www.qdma.com
Rocky Mountain Elk Foundation
http://www.rmef.org/home
Safari Club International
http://www.scifirstforhunters.org
The Sierra Club
http://www.sierraclub.org
Smithsonian Conservation Biology Institute
http://www.nationalzoo.si.edu
Smithsonian National Museum of Natural History
http://www.mnh.si.edu

U.S. Department of Transportation, Federal Highway Administration
http://www.fhwa.dot.gov/index.html
WAZA (World Association of Zoos and Aquariums)
http://www.waza.org/en/site/home
Wildlife Conservation Society
http://www.wcs.org
The Wildlife Society
http://joomla.wildlife.org
WWF (World Wildlife Fund)
http://wwf.panda.org

Bibliography

Allen, M. 1983. *Animals in American Literature*. University of Illinois Press, Urbana.

Apollonio, M., R. Andersen, and R. Putman, eds. 2010. *European Ungulates and Their Management in the 21st Century*. Cambridge University Press, Cambridge.

Aung, M., W. J. McShea, S. Htung, A. Than, et al. 2001. Ecology and social organization of a tropical deer (*Cervus eldi thamin*). *Journal of Mammalogy* 82:836–847.

Bartos, L., B. Fricova, J. Bartosova-Vichova, J. Panamá, et al. 2007. Estimation of the probability of fighting in fallow deer (*Dama dama*) during the rut. *Aggressive Behavior* 33:196–199.

Beck, B. B. and C. Wemmer, eds. *The Biology and Management of an Extinct Species: Père David's Deer*. Noyes Publications, Park Ridge, NJ.

Bildstein, K. L. 1983. Why white-tailed deer flag their tails. *American Naturalist* 121:709–715.

Black-Decima, P. and M. Santana. 2011. Olfactory communication and counter-marking in brown brocket deer, *Mazama gouazoubira*. *Acta Theriologica* 56: 179–187.

Brodie, J. F. and W. Y. Brockelman. 2009. Bed site selection of red muntjac (*Muntiacus muntjak*) and sambar (*Rusa unicolor*) in a tropical seasonal forest. *Ecological Research* 24:1251–1256.

Brown, R. D., ed. 1983. *Antler Development in Cervidae*. Texas A&M University Press, Kingsville.

Brown, R. D., ed. 1992. *The Biology of Deer*. Springer-Verlag, New York.

Bubenik, G. A. and A. Bubenik, eds. 1990. *Horns, Pronghorns, and Antlers*. Springer-Verlag, New York.

Butler, M. J., A. P. Teaschner, W. B. Ballard, and B. K. McGee. 2005. Commentary: Wildlife ranching in North America—arguments, issues, and perspectives. *Wildlife Society Bulletin* 33:381–389.

Cain, J. R., III, P. R. Krausman, S. S. Rosenstock, and J. C. Turner. 2006. Mechanisms of thermoregulation and water balance in desert ungulates. *Wildlife Society Bulletin* 34:570–581.

Carson, A. E. 2005. The hunted stag and the beheaded king. *Studies in English Literature* 45:537–556.

Chapman, D. and N. Chapman. 1975. *Fallow Deer: Their History, Distribution, and Biology*. Terence Dalton, Suffolk, England.

Chapman, N. G., W. A. B. Brown, and P. Rothery. 2005. Assessing the age of Reeves' muntjac (*Muntiacus reevesi*) by scoring wear of the mandibular molars. *Journal of Zoology* 267:223–247.

Chapman, N. G., K. Claydon, M. Claydon, and S. Harris. 1985. Distribution and habitat selection of muntjac and other species of deer in a coniferous forest. *Acta Theriologica* 30:287–303.

Clutton-Brock, T. H. 1982. The function of antlers. *Behaviour* 79:108–125.

Clutton-Brock, T. H., S. D. Albon, and P. H. Harvey. 1980. Antlers, body size, and breeding group size in the Cervidae. *Nature* 285:565–567.

Clutton-Brock, T. H., F. E. Guinness, and S. D. Albon. 1982. *Red Deer: Behavior and Ecology of Two Sexes*. University of Chicago Press, Chicago.

Conner, M. M., M. R. Ebinger, J. A. Blanchong, and P. C. Cross. 2008. Infectious disease in cervids of North America. *Annals of the New York Academy of Sciences* 1134:146–172.

D'Angelo, G. J., A. R. DeChicchis, D. A. Osborn, G. R. Gallagher, R. J. Warren, and K. V. Miller. 2006. Hearing range of white-tailed deer as determined by auditory brainstem response. *Journal of Wildlife Management* 71:1238–1242.

Demaria, M. R., W. J. McShea, K. Koy, and N. O. Maceira. 2004. Pampas deer conservation with respect to habitat loss and protected area considerations in San Luis, Argentina. *Biological Conservation* 115:121–130.

Døving, K. B. and D. Trotier. 1998. Structure and function of the vomeronasal organ. *Journal of Experimental Biology* 201:2913–2925.

Duarte, J. M. B. and S. Gonzalez, eds. 2010. *Neotropical Cervidology: Biology and Medicine of Latin American Deer*. Funep, Jaboticabal, Brazil, and IUCN [International Union for Conservation of Nature], Gland, Switzerland.

Dubost, G., F. Charron, A. Courcoul, and A. Rodier. 2011. Social organization in the Chinese water deer, *Hydropotes inermis*. *Acta Theriologica* 56:189–198.

du Toit, J. T. 2008. Standardizing the data on wildlife-vehicle collisions. *Human-Wildlife Conflicts* 2:5–6.

Eisenberg, J. F. 2000. The contemporary Cervidae of Central and South America. Pages 189–202 *in* E. S. Vrba and G. B. Schaller, eds. *Antelopes, Deer, and Relatives: Fossil Record, Behavioral Ecology, Systematics, and Conservation*. Yale University Press, New Haven, CT.

Epstein, M. B., G. A. Feldhamer, and R. L. Joyner. 1983. Predation on white-tailed deer fawns by bobcats, foxes, and alligators: Predator assessment. *Proceedings, Annual Conference of the Southeastern Association of Fish and Wildlife Agencies* 37: 161–172.

Feighny, J. A., K. E. Williamson, and J. A. Clarke. 2006. North American elk bugle vocalizations: Male and female bugle call structure and content. *Journal of Mammalogy* 87:1072–1077.

Feldhamer, G. A. 1980. Sika deer, *Cervus nippon*. *Mammalian Species* 128:1–7.

Feldhamer, G. A. 2002. Acorns and white-tailed deer (*Odocoileus virginianus*): Interrelationships in forest ecosystems. Pages 215–223 *in* W. J. McShea and W. M. Healy, eds. *The Ecology and Management of Oaks for Wildlife*. Johns Hopkins University Press, Baltimore.

Feldhamer, G. A. and S. Demarais. 2008. Free-ranging and confined sika deer in North America: Current status, biology, and management. Pages 615–641 *in* D. R. McCullough, S. Takatsuki, and K. Kaji, eds. *Sika Deer: Biology and Management of Native and Introduced Populations*. Tokyo, Springer Japan.

Feldhamer, G. A., J. E. Gates, D. M. Harman, A. Loranger, and K. R. Dixon. 1986. Effects of interstate highway fencing on white-tailed deer activity. *Journal of Wildlife Management* 50:497–503.

Feldhamer, G. A., T. P. Kilbane, and D. W. Sharp. 1989. Effect of cumulative winter climate on acorn yield and body weight of immature deer. *Journal of Wildlife Management* 53:292–295.

Feldhamer, G. A. and M. A. Marcus. 1994. Reproductive performance of female sika deer in Maryland. *Journal of Wildlife Management* 58:670–673.

Feldhamer, G. A., D. W. Sharp, and T. Davin. 1992. Acorn yield and yearling white-tailed deer on Land Between the Lakes, Tennessee. *Journal of the Tennessee Academy of Science* 67:46–48.

Feldhamer, G. A., J. R. Stauffer, and J. A. Chapman. 1985. Body morphology and weight relationships of sika deer (*Cervus nippon*) in Maryland. *Zeitschrift für Säugetierkunde* 50:88–106.

Feldhamer, G. A., B. C. Thompson, and J. A. Chapman, eds. 2003. *Wild Mammals of North America: Biology, Management, and Conservation*. Johns Hopkins University Press, Baltimore.

Fennessy, P. F. and K. R. Drew, eds. 1985. *Biology of Deer Production*. Royal Society of New Zealand Bulletin No. 22. Royal Society of New Zealand, Wellington.

Fletcher, J. D., W. J. McShea, L. A. Shipley, and D. Shumway. 2001. Use of common forest forbs to measure browsing pressure by white-tailed deer (*Odocoileus virginianus* Zimmerman) in Virginia, USA. *Natural Areas Journal* 21:172–176.

Franzmann, A. W. and C. C. Schwartz, eds. 1998. *Ecology and Management of North American Moose*. Smithsonian Institution Press, Washington, DC.

Frid, A. 1994. Observations on habitat use and social organization of a huemul (*Hippocamelus bisulcus*) coastal population in Chile. *Biological Conservation* 67:13–20.

Frid, A. 2001. Habitat use by endangered huemul (*Hippocamelus bisulcus*): Cattle, snow, and the problem of multiple causes. *Biological Conservation* 100:261–267

Geist, V. 1998. *Deer of the World: Their Evolution, Behavior, and Ecology*. Stackpole Books, Mechanicsburg, PA.

Geist, V. and M. Bayer. 1988. Sexual dimorphism in the Cervidae and its relation to habitat. *Journal of Zoology* (London) 214:45–53.

Geist, V. and P. T. Bromley. 1978. Why deer shed antlers. *Zeitschrift für Säugetierkunde* 43:223–232.

Geist, V. and F. Walther, eds. 1974. *The Behaviour of Ungulates and Its Relation to Management*, 2 vols. IUCN Publication, n.s., No. 24. International Union for Conservation of Nature, Morges, Switzerland.

Gentry, A. W. 1994. The Miocene differentiation of Old World Pecora (Mammalia). *Historical Biology* 7:115–158.

Gentry, A. W. 2000. The ruminant radiation. Pages 11–25 *in* E. S. Vrba and G. B. Schaller, eds. *Antelopes, Deer, and Relatives: Fossil Record, Behavioral Ecology, Systematics, and Conservation*. Yale University Press, New Haven, CT.

González, S., F. Álvarez-Valin, and J. E. Maldonado. 2002. Morphometric differentiation of endangered pampas deer (*Ozotoceros bezoarticus*), with description of new subspecies from Uruguay. *Journal of Mammalogy* 83:1127–1140.

González, S., J. E. Maldonado, J. Ortega, A. C. Talarico, et al. 2009. Identification of the endangered small red brocket deer (*Mazama bororo*) using noninvasive genetic techniques (Mammalia: Cervidae). *Molecular Ecology Resources* 9:754–758.

Goss, R. J. 1983. *Deer Antlers: Regeneration, Function, and Evolution*. Academic Press, New York.

Gould, S. J. 1974. The origin and function of "bizarre" structures: Antler size and skull size in the "Irish elk" *Megaloceros giganteus*. *Evolution* 28:221–231.

Grassman, B. T. and E. C. Hellgren. 1993. Phosphorus nutrition in white-tailed deer: Nutrient balance, physiological responses, and antler growth. *Ecology* 74:2279–2296.

Groves, C. P. 2007. Family Cervidae. Pages 249–256 *in* D. R. Prothero and S. E. Foss, eds. *The Evolution of Artiodactyls*. John Hopkins University Press, Baltimore.

Halls, L. K., ed. 1984. *White-Tailed Deer: Ecology and Management*. Stackpole Books, Harrisburg, PA.

Hassig, D. 1997. Marginal bestiaries. Pages 171–188 *in* L. A. J. R. Houwen, ed. *Animals and the Symbolic in Mediaeval Art and Literature*. Egbert Forsten, Groningen, the Netherlands.

Heckel, C. D., N. A. Bourg, W. J. McShea, and S. Kalisz. 2010. Nonconsumptive effects of a generalist ungulate herbivore drive decline of unpalatable forest herbs. *Ecology* 91:319–326.

Heffelfinger, J. 2006. *Deer of the Southwest: A Complete Guide to the Natural History, Biology, and Management of Southwestern Mule Deer and White-Tailed Deer*. Texas A&M University Press, College Station.

Hemami, M. R., A. R. Watkinson, and P. M. Dolman. 2005. Population densities and habitat associations of introduced muntjac *Muntiacus reevesi* and native roe deer *Capreolus capreolus* in a lowland pine forest. *Forest Ecology and Management* 215:224–238.

Henshaw, J. 1969. Antlers—Bones of contention. *Nature* 244:1036–1037.

Hirth, D. H. 1977. *Social Behavior of White-Tailed Deer in Relation to Habitat*. Wildlife Monographs No. 53. Wildlife Society, [Washington, DC].

Hirth, D. H. and D. R. McCullough. 1977. Evolution of alarm signals in ungulates with specific references to white-tailed deer. *American Naturalist* 111:31–42.

Hofmann, R. R. 1985. Digestive physiology of the deer—Their morphophysiological specialisation and adaptation. Pages 393–407 *in* P. F. Fennessy and K. R. Drew, eds. *Biology of Deer Production*. Royal Society of New Zealand Bulletin No. 22. Royal Society of New Zealand, Wellington.

Houwen, L. A. J. R., ed. *Animals and the Symbolic in Mediaeval Art and Literature*. Egbert Forsten Publishing, Groningen, the Netherlands.

Hu, H. J., and Z. G. Jiang. 2002. Trial release of Père David's deer *Elaphurus davidianus* in the Dafeng Reserve, China. *Oryx* 36:196–199.

Hu, J., S. G. Fang, and Z. H. Wan. 2006. Genetic diversity of Chinese water deer (*Hydropotes inermis inermis*): Implications for conservation. *Biochemical Genetics* 44:161–172.

Hundertmark, K. J., G. F. Shields, I. G. Udina, R. T. Bowyer, et al. 2002. Mitochondrial phylogeography of moose (*Alces alces*): Late Pleistocene divergence and population expansion. *Molecular Phylogenetics and Evolution* 22:375–387.

Huston, J. E., B. S. Rector, W. C. Ellis, and M. L. Allen. 1986. Dynamics of digestion in cattle, sheep, goats, and deer. *Journal of Animal Science* 62:208–215.

Illius, A. W. 1997. Physiological adaptation in savanna ungulates. *Proceedings of the Nutrition Society* 56:1041–1048.

Jackson, J. E. 1986. Antler cycle in pampas deer (*Ozotoceros bezoarticus*) from San Luise, Argentina. *Journal of Mammalogy* 67:173–176.

James, J., U. Ramakrishnan, and A. Datta. 2008. Molecular evidence for the occurrence of the leaf deer *Muntiacus putaoensis* in Arunachal Pradesh, north-east India. *Conservation Genetics* 9:927–931.

Jarman, P. J. 2000. Dimorphism in social Artiodactyla: Selection upon females. Pages 171–179 *in* E. S. Vrba and G. B. Schaller, eds. *Antelopes, Deer, and Relatives: Fossil Record, Behavioral Ecology, Systematics, and Conservation*. Yale University Press, New Haven, CT.

Jiang, Z., S. Hamasaki, S. Takatsuki, M. Kishimoto, and M. Kitahara. 2009. Seasonal and sexual variation in the diet and gastrointestinal features of the sika deer in western Japan: Implications for the feeding strategy. *Zoological Science* 26:691–697.

Keiper, R. R. 1985. Are sika deer responsible for the decline of white-tailed deer on Assateague Island, Maryland? *Wildlife Society Bulletin* 13:14–146.

Kiddie, D. G. 1962. The sika deer (*Cervus nippon*) in New Zealand. *New Zealand Forest Service, Information Series* 44:1–35.

Kong, Y. C. and P. P. H. But. 1985. Deer—the ultimate medicinal animal (antler and deer parts in medicine). Pages 311–324 *in* P. F. Fennessy and K. R. Drew, eds. *Biology of Deer Production*. Royal Society of New Zealand Bulletin No. 22. Royal Society of New Zealand, Wellington.

Koy, K., W. J. McShea, P. Leimgruber, B. N. Haack, et al. 2005. Percentage canopy cover—Using Landsat imagery to delineate habitat for Myanmar's endangered Eld's deer (*Cervus eldi*). *Animal Conservation* 8:289–296.

Kuiters, A. T., G. M. J. Mohren, and S. E. Van Wieren. 1996. Ungulates in temperate forest ecosystems. *Forest Ecology and Management* 88:1–5.

LaGory, K. A. 1986. Habitat, group size, and the behaviour of white-tailed deer. *Behaviour* 98:168–179.

Lever, C. 1985. *Naturalized Mammals of the World*. Longman, New York.

Lima, S. L., N. C. Rattenborg, J. A. Lesku, and C. J. Amlaner. 2005. Sleeping under the risk of predation. *Animal Behaviour* 70:723–736.

Lister, A. M. 1994. The evolution of the giant *Megaloceros giganteus* (Blumbach). *Zoological Journal of the Linnean Society* 112:65–100.

Lowe, V. P. W. and A. S. Gardiner. 1989. Are the New and Old World wapitis conspecific with red deer (*Cervus elaphus*)? *Journal of Zoology* (London) 218:51–58.

Lydekker, R. 1898. *The Deer of All Lands: A History of the Family Cervidae Living and Extinct*. Rowland Ward, London.

Mayze, R. J. and G. I. Moore. 1990. *The Hog Deer*. Australian Deer Research Foundation, Croydon, Australia.

McCullough, D. R. 1979. *The George Reserve Deer Herd: Population Ecology of a K-Selected Species*. University of Michigan Press, Ann Arbor.

McCullough, D. R., S. Takatsuki, and K. Kaji, eds. 2009. *Sika Deer: Biology and Management of Native and Introduced Populations*. Tokyo, Springer Japan.

McElligot, A. G., V. Mattiangeli, S. Mattiello, M. Verga, et al. 1998. Fighting tactics of fallow bucks (*Dama dama*, Cervidae): Reducing the risks of serious conflict. *Ethology* 104:789–803.

McShea, W. J. 2000. The influence of acorn crops on annual variation in rodent and bird populations. *Ecology* 81:228–238.

McShea, W. J. 2005. Forest ecosystems without carnivores: When ungulates rule the world. Pages 138–153 *in* J. C. Ray, K. Redford, R. S. Stenbeck, and J. Berger, eds. *Large Carnivores and the Conservation of Biodiversity*. Island Press, Washington, DC.

McShea, W. J., M. Aung, D. Poszig, C. Wemmer, et al. 2001. Forage, habitat use, and sexual segregation by a tropical deer (*Cervus eldi thamin*) in a dipterocarp forest. *Journal of Mammalogy* 82:848–857.

McShea, W. J., K. Koy, T. Clements, A. Johnson, et al. 2005. Finding a needle in the haystack: Regional analysis of suitable Eld's deer (*Cervus eldi*) forest in Southeast Asia. *Biological Conservation* 125:101–111.

McShea, W. J., S. L. Monfort, S. Hakim, J. Kirkpatrick, I. Liu, J. W. Turner, L. Chassy, and L. Munson. 1997. The effect of immunocontraception on the behavior and reproduction of white-tailed deer. *Journal of Wildlife Management* 61:560–569.

McShea, W. J. and J. H. Rappole. 1997. The science and politics of managing deer within a protected area. *Wildlife Society Bulletin* 25:443–446.
McShea, W. J. and J. H. Rappole. 2000. Managing the abundance and diversity of breeding bird populations through manipulation of deer populations. *Conservation Biology* 14:1161–1170.
McShea, W. J. and G. Schwede. 1993. Variable acorn crops and the response of white-tailed deer and other mast consumers. *Journal of Mammalogy* 74:999–1006.
McShea, W. J., C. M. Stewart, L. J. Kearns, S. Liccioli, et al. 2008. Factors affecting autumn deer-vehicle collisions in a rural Virginia county. *Human-Wildlife Conflicts* 2:110–121.
McShea, W. J., H. B. Underwood, and J. H. Rappole, eds. 1997. *The Science of Overabundance: Deer Ecology and Population Management*. Smithsonian Institution Press, Washington, DC.
Messmer, T. A. and D. R. Messmer. 2008. Deer-vehicle collision statistics and mitigation information: Online sources. *Human-Wildlife Conflicts* 2:131–135.
Miller, K. V. and R. L. Marchinton, eds. 1995. *Quality Whitetails: The Why and How of Quality Deer Management*. Stackpole Books, Mechanicsburg, PA.
Milner, J. M., C. Bonenfant, A. Mysterud, J.-M. Gaillard, S. Csányi, and N. C. Stenseth. 2006. Temporal and spatial development of red deer harvesting in Europe: Biological and cultural factors. *Journal of Applied Ecology* 43:721–734.
Monfort, S. L., J. L. Brown, M. Bush, T. C. Wood, C. Wemmer, A. Vargas, L. R. Williamson, R. J. Montali, and D. E. Wildt. 1993. Circannual inter-relationships among reproductive hormones, gross morphometry, behaviour, ejaculate characteristics, and testicular histology in Eld's deer stags (*Cervus eldi thamin*). *Journal of Reproduction and Fertility* 98:471–480.
Monteith, K. L., V. C. Bleich, T. R. Stephenson, B. M. Pierce, et al. 2011. Timing of seasonal migration in mule deer: Effects of climate, plant phenology, and life-history characteristics. *Ecosphere* 2:1–34.
Mullan, J. M., G. A. Feldhamer, and D. Morton. 1988. Reproductive characteristics of female sika deer in Maryland and Virginia. *Journal of Mammalogy* 69:388–389.
Odden, M., P. Wegge, and T. Storaas. 2005. Hog deer *Axis porcinus* need threatened tallgrass floodplains: A study of habitat selection in lowland Nepal. *Animal Conservation* 8:99–104.
Olesen, C. R. and P. Madsen. 2008. The impact of roe deer (*Capreolus capreolus*), seedbed, light, and seed fall on natural beech (*Fagus sylvatica*) regeneration. *Forest Ecology and Management* 255:3962–3972.
Olsen, K. 1997. Animated ships in Old English and Old Norse poetry. Pages 53–66 *in* L. A. J. R. Houwen, ed. *Animals and the Symbolic in Mediaeval Art and Literature*. Egbert Forsten, Groningen, the Netherlands.
Otaishi, N. and H. I. Sheng, eds. *Deer of China: Biology and Management*. Elsevier, Amsterdam.
Perkins, D. 1998. Wordsworth and the polemic against hunting: "Hart-Leap Well." *Nineteenth-Century Literature* 52:421–445.
Pitra, C., J. Fickel, E. Meijaard, and P. C. Groves. 2004. Evolution and phylogeny of Old World deer. *Molecular Phylogenetics and Evolution* 33:880–895.
Prins, R. A. and M. J. H. Geelen. 1971. Rumen characteristics of red deer, fallow deer, and roe deer. *Journal of Wildlife Management* 35:673–680.

Prothero, D. R. and R. M. Schoch. 2002. *Horns, Tusks, and Flippers: The Evolution of Hoofed Mammals* [see especially chapter 2 ("Cloven hoofs") and chapter 4 ("Where deer and antelope play")]. Johns Hopkins University Press, Baltimore.

Putman, R. J. 1988. *The Natural History of Deer*. Cornell University Press, Ithaca, NY.

Putman, R. J. and B. W. Staines. 2004. Supplementary winter feeding of wild red deer *Cervus elaphus* in Europe and North America: Justifications, feeding practice, and effectiveness. *Mammal Review* 34:285–306.

Rice, P. R. and D. C. Church. 1974. Taste responses of deer to browse extracts, organic acids, and others. *Journal of Wildlife Management* 38:830–836.

Richardson, L. W., H. A. Jacobson, R. J. Muncy, and C. J. Perkins. 1983. Acoustics of white-tailed deer (*Odocoileus virginianus*). *Journal of Mammalogy* 64:245–252.

Rivero, K., D. I. Rumiz, and A. B. Taber. 2005. Differential habitat use by two sympatric brocket deer species (*Mazama americana* and *M. gouazoubira*) in a seasonal Chiquitano forest of Bolivia. *Mammalia* 69:169–183.

Rue, L. L. 2003. *The Encyclopedia of Deer*. Voyageur Press, Stillwater, MN.

Ryg, M. 1986. Physiological control of growth, reproduction, and lactation in deer. *Rangifer* 1:261–266.

Sawyer, T. G., R. L. Marchinton, and K. V. Miller. 1989. Response of female white-tailed deer to scrapes and antler rubs. *Journal of Mammalogy* 70:431–433.

Schaller, G. B. and E. S. Vrba. 1996. Description of the giant muntjac (*Megamuntjacus vuquangensis*) in Laos. *Journal of Mammalogy* 77:675–683.

Schmidt, J. L. and D. C. Gilbert, eds. 1978. *Big Game of North America*. Stackpole Books, Harrisburg, PA.

Schwede, G., H. Hendrichs, and W. McShea. 1993. Social and spatial organization of female white-tailed deer, *Odocoileus virginianus*, during the fawning season. *Animal Behaviour* 45:1007–1017.

Scott, W. B. 1885. *Cervalces americanus*, a fossil moose, or elk, from the quaternary of New Jersey. *Proceedings of the Academy of Natural Sciences of Philadelphia* 37:181–202.

Sheffield, S., R. P. Morgan, G. A. Feldhamer, and D. M. Harman. 1985. Genetic differences in white-tailed deer populations in western Maryland. *Journal of Mammalogy* 66:243–255.

Silva-Rodríguez, E. A., C. Verdugo, O. A. Aleuy, J. G. Sanderson, et al. 2010. Evaluating mortality sources for the vulnerable pudu *Pudu puda* in Chile: Implications for the conservation of a threatened deer. *Oryx* 44:97–103.

Smith, M. H., W. V. Brannan, R. L. Marchington, P. E. Johns, et al. 1986. Genetic and morphological comparison of red brocket, brown brocket, and white-tailed deer. *Journal of Mammalogy* 67:103–111.

Sorin, A. B. 2004. Paternity assignment for white-tailed deer (*Odocoileus virginianus*): Mating across age classes and multiple paternity. *Journal of Mammalogy* 85: 356–362.

Stallings, J. R. 1986. Notes on the reproductive biology of the grey brocket deer (*Mazama gouazoubira*) in Paraguay. *Journal of Mammalogy* 67:172–175.

Stankowich, T. and R. G. Coss. 2008. Alarm walking in Columbian black-tailed deer: Its characterization and possible antipredatory signaling functions. *Journal of Mammalogy* 89:636–645.

Stewart, C. M., W. J. McShea, and B. P. Piccolo. 2007. The impact of white-tailed deer on agricultural landscapes in 3 national historical parks in Maryland. *Journal of Wildlife Management* 71:1525–1530.

Stonehouse, B. 1968. Thermoregulatory function of growing antlers. *Nature* 218: 870–872.

Taylor, W. P., ed. 1956. *The Deer of North America*. Stackpole Books, Harrisburg, PA.

Thomas, J. W. and D. E. Toweill, eds. 1982. *The Elk of North America: Ecology and Management*. Stackpole Books, Harrisburg, PA.

Thomas, K. 1983. *Man and the Natural World: Changing Attitudes in England, 1500–1800*. Allen Lane, London.

Tomas, W. M. and S. M. Salis. 2000. Diet of the marsh deer (*Blastocerus dichotomus*) in the Pantanal wetland, Brazil. *Studies on Neotropical Fauna and Environment* 35:165–172.

Vannoni, E. and A. G. McElligott. 2007. Individual acoustic variation in fallow deer (*Dama dama*) common and harsh groans: A source-filter theory perspective. *Ethology* 113:223–234.

Vanpe, C., P. Kjellander, J. M. Gaillard, J. F. Cosson, et al. 2009. Multiple paternity occurs with low frequency in the territorial roe deer, *Capreolus capreolus*. *Biological Journal of the Linnean Society* 97:128–139.

VerCauteren, K. C. and M. J. Pipas. 2003. A review of color vision in white-tailed deer. *Wildlife Society Bulletin* 31:684–691.

Verme, L. J. and J. L. Ozoga. 1980. Effects of diet on growth and lipogenesis in deer fawns. *Journal of Wildlife Management* 44: 315–324.

Wallmo, O. C., ed. 1981. *Mule and Black-Tailed Deer of North America*. University of Nebraska Press, Lincoln.

Webb, S. D. 2000. Evolutionary history of New World Cervidae. Pages 38–64 *in* E. S. Vrba and G. B. Schaller, eds. *Antelopes, Deer, and Relatives: Fossil Record, Behavioral Ecology, Systematics, and Conservation*. Yale University Press, New Haven, CT.

Wemmer, C. M., ed. 1987. *Biology and Management of the Cervidae*. Smithsonian Institution Press, Washington, DC.

Whitehead, G. K. 1972. *Deer of the World*. Constable, London.

Whitehead, G. K. 1982. *Hunting and Stalking Deer Throughout the World*. Batsford, London.

Xie, Z., K. I. O'Rourke, Z. Dong, A. L. Jenny, et al. 2006. Chronic wasting disease of elk and deer and Creutzfeldt-Jakob disease: Comparative analysis of the scrapie prion protein. *Journal of Biological Chemistry* 281:4199–4206.

Yahner, R. H. 1980. Barking in a primitive ungulate, *Muntiacus reevesi*: Function and adaptiveness. *American Naturalist* 116:157–177.

Zepelin, H. 2000. Mammalian sleep. Pages 82–92 *in* M. H. Kryger, T. Roth, and W. C. Dement, eds. *Principles and Practice of Sleep Medicine*. Saunders, Philadelphia.

Ziolkowski, J. M. 1997. Literary genre and animal symbolism. Pages 1–23 *in* L. A. J. R. Houwen, ed. *Animals and the Symbolic in Mediaeval Art and Literature*. Egbert Forsten, Groningen, the Netherlands.

Index

Page references in italics refer to illustrations. Page numbers followed by *t* refer to tables.